W0111360

Coding the Arduino

Building Fun Programs, Games, and Electronic Projects

Bob Dukish

Apress®

Coding the Arduino: Building Fun Programs, Games, and Electronic Projects

Bob Dukish
Canfield, Ohio, USA

ISBN-13 (pbk): 978-1-4842-3509-6   ISBN-13 (electronic): 978-1-4842-3510-2
https://doi.org/10.1007/978-1-4842-3510-2

Library of Congress Control Number: 2018945863

Managing Director, Apress Media LLC: Welmoed Spahr
Acquisitions Editor: Natlie Pao
Development Editor: James Markham
Coordinating Editor: Jessica Vakili

Cover designed by eStudioCalamar

Cover image designed by Freepik (www.freepik.com)

Distributed to the book trade worldwide by Springer Science+Business Media New York, 233 Spring Street, 6th Floor, New York, NY 10013. Phone 1-800-SPRINGER, fax (201) 348-4505, e-mail orders-ny@springer-sbm.com, or visit www.springeronline.com. Apress Media, LLC is a California LLC and the sole member (owner) is Springer Science + Business Media Finance Inc (SSBM Finance Inc). SSBM Finance Inc is a **Delaware** corporation.

For information on translations, please e-mail rights@apress.com, or visit http://www.apress.com/rights-permissions.

Apress titles may be purchased in bulk for academic, corporate, or promotional use. eBook versions and licenses are also available for most titles. For more information, reference our Print and eBook Bulk Sales web page at http://www.apress.com/bulk-sales.

Any source code or other supplementary material referenced by the author in this book is available to readers on GitHub via the book's product page, located at www.apress.com/978-1-4842-3509-6. For more detailed information, please visit http://www.apress.com/source-code.

Printed on acid-free paper

Table of Contents

About the Author

Bob Dukish has been working in the field of computers and electronics for over 35 years. He served in the military, worked as an electronic component engineer, ran a corporation, and taught engineering at both the high school and college levels. He has two associate's degrees in technology, a bachelor's degree in physics from Syracuse University, and master's degrees from both Kent State University and Rensselaer Polytechnic Institute. His last master's degree was earned at the age of 54, and he considers himself to be a lifelong learner.

About the Technical Reviewers

Dave Brett started his electronics career in the U.S. Air Force as an instructor in the Radar School at Keesler AFB. He went on to work as a technician for the Ohio State University, and as a 2-way radio technician for MSS. Dave taught electronics for many years at ITT Technical Institute in Youngstown Ohio, and is now is an Instructor at the Pittsburgh Institute of Aeronautics. He graduated from Youngstown State University with a master's degree in Education and is certified by the Electronics Technician Association, CompTia, and the Society of Broadcast Engineers. Dave is an avid Amateur Radio enthusiast and participates in the Amateur Radio Emergency Service.

Mark Furman, MBA is a systems engineer, author, teacher, and entrepreneur. For the last 18 years he has worked in the information technology field with a focus on Linux-based systems and programming in Python, working for a range of companies including Host Gator, Interland, Suntrust Bank, AT&T, and Winn-Dixie. Recently he has been focusing his career on the maker movement and has launched Tech Forge (techforge.org). He holds a master's of business administration degree with a focus on business intelligence from Ohio University. You can follow him on Twitter at @mfurman.

Warning

Electrical circuits and components may contain lethal voltages even when disconnected. Do not attempt to test, modify, or repair electrical equipment. Hazardous voltages might be present, and even low voltages can produce high currents that can cause severe burns. Care must also be taken, as some Arduino boards have exposed solder connections that could come in contact with conductive materials and cause a short circuit.

Introduction

Communication and Creativity

Life-forms on our planet are biologically programmed through evolution to be interested in their surroundings for self-preservation, but some go a step further. There is a popular expression: "Curiosity killed the cat." Our human species is extraordinarily inquisitive as well (although not equal to the cat), but it is our curiosity coupled with communication and creativity that has propelled humankind to become the dominant species on the planet. What is truly special about the human race is our ability to discover, retain, convey, and most important, synthesize new concepts. The communal knowledge that we amass allows us as a species to learn from past experiences, and we use our creativity to develop entirely new ideas. This has brought about modern technological marvels such as telephone, television, computers, and all of the other items ubiquitous in our modern lifestyle. Humankind's insatiable need to be linked together with others and communicate information builds a database of knowledge where creative thought can then be applied to synthesize new concepts. This is undeniably how the exponential growth in technological advancement has occurred. Paraphrasing Sir Isaac Newton, we stand "on the shoulders of giants."

We can extrapolate back to prehistoric times and make an educated assumption that knowledge was shared in early societies by individuals patterning after others within a group even before spoken or written language was developed. As time progressed and history developed, we know through writings that early humans sought to satisfy their curiosity and used creative thought to make sense of the world around them.

Most early civilizations imagined that mystical entities brought about order to the surrounding world, and the dichotomy of good and bad was explained as being the intent of either benevolent or malevolent deities. Early Greek mythology gave fanciful explanations of the world by looking up to the unreachable stars overhead and associating their patterns with supernatural concepts. Later, their civilization provided humanity with the beginnings of science from enlightened explanations of the physical world deduced through logical reasoning. The early Greek scholars' explanation of indirectly observable phenomenon such as electricity provides us a working knowledge that is somewhat still in use to this day. Very quickly, in the grand scheme of things, humankind went from thinking everything was magical and out of human control to a basic understanding of the atom as being an indivisible building block of the chemical elements that make up the universe.

It seems that we have now come to the point where there is an exponential growth function of the advancement of knowledge leading to great leaps in both science and technology that are almost explosive! Atoms are building blocks of matter, just as the early Greeks thought, but late nineteenth- and early twentieth-century science had discovered that atoms were constructed of a collection of three subatomic particles: electrons, protons, and neutrons. Thanks to mathematicians, particle physicists, and supercolliders, we now know that the protons and neutrons are made up of even smaller subatomic particles called quarks, to which physicists have given fanciful names in identifying different varieties such as top, bottom, up, down, charm, and strange.

With our wondrous machines actually able to peer inside of individual atoms, and through painstaking theoretical work in mathematics and science, humankind has achieved such a detailed understanding of the physical structure of matter and the interactions of energy, we now know that there are more than 100 subatomic particles dealing with matter and forces. The universe is just as beautifully complex as it is immense. Beyond narrow religious views, nationalistic fervor, race, and socioeconomic status, the grandeur of the universe should resonate with us and unite all of humankind. Unfortunately, parochial systems persist, and we have amassed the knowledge and technology to obliterate the planet we live on. Several nations across the globe have a hairpin trigger on nuclear devices that could purposefully destroy our entire civilization. Perhaps the reason that we have not been able to eavesdrop on communication signals emanating from civilizations orbiting other stars is that they are either too young or have gotten to the point at which we are now and have developed nuclear weapons and destroyed themselves. Let's hope for the best for them, and for ourselves.

About This Book

This book is intended for someone new to computer coding and electronics technology. It contains four sections. The first provides a background on electronic components and circuits. We then begin writing game code for an Arduino development board using a subset of the popular programming language called C++. In the third section, we build electronic game and communications projects, and modify some of the code presented in previous chapters to operate the devices. The fourth section expands on the functionality of some of the programs presented in previous chapters and challenges the reader with capstone projects.

As we present programs throughout the text, and later make modifications to perform additional functions, we will generally rewrite the original code and highlight new code placed into the more functional programs. At the end of each chapter, there are review questions that allow the reader an opportunity to test his or her comprehension of the material. Additionally, coding projects will be described where the program code that is presented can be modified, or in which two or more of the sample programs can be used to synthesize a new program as the solution to the problem that is presented. Answers to both the review questions and solution help to the coding projects appear in the Appendix. Additionally, the Appendix contains information about the use of Arduino libraries that simplify program coding.

There are many different ways to code a program, just as there are many different routes that can be taken on a trip between two points on the globe. The final objective in traveling is to arrive at an intended destination. I consider the learning process to be like a trek along an infinite pathway, and many of the examples in this text take what might be termed the scenic route to discover new and interesting things along the way. This helps make the learning experience more immersive, just as if one were on vacation and able to spend additional time exploring unknown areas of the world to discover new things. It is also hoped that the adventurous learner will experiment with the programs by coding modifications to the projects as they are presented.

Arduino boards are available from the official Arduino web site at `www.arduino.cc`, and from many electronics suppliers. Inexpensive parts kits containing resistors, light-emitting diodes (LEDs), integrated circuits (ICs), and other items discussed in this text are available through a number of sources. Links to parts outlets and some of the lengthier code examples can be obtained as a free download from the author's official web site at `www.dukish.com`.

Acknowledgments

A very intelligent gentleman who worked as a professional house painter offered thoughtful advice when I complained that a job was so massive that it would take "forever" to complete. His response was to not look at the overall project, but to only concentrate on one section at a given time. That advice rings true in every aspect of life, and especially in complex areas like computer hardware design and software programming. What at first glance might seem insurmountably difficult to comprehend can indeed be conquered by having laser-like focus and taking things one step at a time. Thank you, Tom Martinko. Thank you to the code reviewers Dave Brett and Mark Furman who tested every line of code for functionality. I would also like to thank my students from the Trumbull Correctional Institution in Ohio and their desire to overcome adversity and achieve success as productive citizens by gaining new employment skills. Finally, I would like to express my gratitude to the great college instructors I was lucky to have had, who helped me understand complex material by not putting tedious and unnecessary roadblocks in the way.

A Note from the Author About Education

Many years ago, I had an excellent experience in the military where it was strongly encouraged that airmen take college courses and work toward a college degree. I attended night classes at both Mohawk Valley Community College and Utica College of Syracuse University, but struggled with what are now termed STEM courses. While struggling in college, I was lucky to have a great physics teacher who suggested the best way to learn complex material was to read and reread the text, as many times as it took to truly understand the concepts. That teacher also had an excellent suggestion on textbook problem solving: "Try working out a problem, and if the answer was incorrect, take a break and later retry solving the same problem." I heeded the advice and my college textbooks were well read, and numerous end-of-chapter problems were worked until the solution was correct and understood. There is much truth in the sayings that patience is a virtue and ignorance is bliss. Now as a teacher, I feel extremely honored to be able to pass that information on to students covering complex material. I also had a great English teacher for my first college writing class. I mentioned to him that I did not remember anything from my high school English classes about verbs, predicates, and pronouns. His advice was, "Forget about all of that, and just write how you speak," only to clean it up and be more formal.

INTRODUCTION

I do not expect to win a writing contest for this book, but I hope it provides some new and interesting information.

> *A doctor can bury his mistakes, but an architect can only advise his clients to plant vines.*

> —Frank Lloyd Wright

Let's build a few programs for fun and worry about the vines later. If you push the wrong button or enter the wrong code, you need not worry; the computer won't blow up! Also, no matter how lengthy, repetitive, or ugly the code that we write in implementing the objectives in this text, we will be successful if the program works and produces the intended result. Before money mattered, I am sure Bill Gates—now the richest man in the world—just had fun playing with computer code to produce simple tasks. Let's have fun and learn new ways of thinking. We can worry about perfecting the code and making money later.

CHAPTER 1

A Background on Technology

The Difference Between Science and Technology

The two words *science* and *technology* are used interchangeably in the everyday world, but the fields are distinguishably different. As a technologist, one should have a profound appreciation of science; however, it is imperative that a technologist not only appreciate and understand general scientific concepts, but also be able to apply them to the everyday world. Essentially, science can be thought of as a body of knowledge with technology being the practical application of that knowledge. To be an effective electrical engineer or technician, for example, it helps to have an understanding of the actual physical theory of materials and electricity, but many times we will take a simplified approach to solve specific problems. To gain an understanding of the reasoning for simplification in problem solving, please refer to Figure 1-1, a drawing of the copper atom.

© Bob Dukish 2018
B. Dukish, *Coding the Arduino*, https://doi.org/10.1007/978-1-4842-3510-2_1

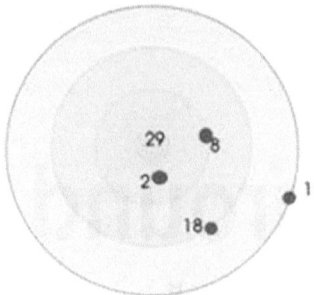

Figure 1-1. *The Bohr model of the copper atom*

The copper atom is composed of 29 protons, each having a positive charge, and located at the center of the atom. Surrounding the protons and uncharged neutrons in the nucleus are 29 negatively charged electrons in several thin spherical clouds located at distances from the center. The location of each cloud is dependent on the energy level of the electrons it contains. Electrons with higher energy levels are located farther away from the center. Like charges repel and unlike charges attract in an inverse square relationship to the distance between charges. In the element copper, there is a single electron called the *valance electron* in the highest energy level, and that electron is loosely bound to the atom because of its distance from the nucleus. The basic original theory of charge, and even the name *electron,* comes about from the work of early Greek scholars more than 2,500 years ago, who theorized about electrostatic interactions between cloth and the substance known as amber. More recently, physicists in the early 1900s helped to refine our basic understanding of the structure of matter. Through studies of the nature of electricity, it is known that in a conductive wire, such as one made of copper, if given an amount of external energy from a power source such as a battery, the electron farthest away from the nucleus can become free, and escape the atom to flow with an organized electric current through the wire, eventually joining an atom farther down the line that has a vacancy, called a *hole,* from the loss of its highest energy electron. Although the movement of each electron, called *drift,* takes a slight amount of time, the effective signal speed through the entire wire occurs at roughly three-fourths the speed of light.

With the preceding explanation, it is possible to have a very good working knowledge of how a conductor works. Please note that materials at the atomic level are actually much more complicated due to recently discovered quantum theory, but we do not need to discuss the subatomic quarks to understand the essential mechanics for electric current flow. Our technological discussion, therefore, relies mainly on the educated guesses of ancient Greek scholars 2,500 years ago, and through the groundbreaking, but now outdated, explanation of the construction of atoms by physicist Niels Bohr in the 1920s, which is enough to give us a simplified working model of matter as it relates to current flow through a conductor. Now, let us go back in time about 200 years, to the days of one of America's greatest scientists, Ben Franklin, who was without an understanding of the atomic theory, for which Bohr was awarded the Nobel Prize in 1922. Ben Franklin used intuition and common sense, and hypothesized that electric flow most probably flowed like water, from a high level, to one that is lower. He felt that, like gravity, the electric force pulled down toward a low point of charge. We now typically refer to this low point as either *ground, neutral,* or *return.*

Many college courses in electrical engineering still use Ben Franklin's conventional current theory to evaluate circuits like the one shown in Figure 1-2, even though Franklin's flow, called *conventional,* is completely backward! Thanks to the work of Bohr and other scientists of the 20th century, we now know that the negative electrons are the current carrier, as the proton is more massive and locked within the nucleus, but we can simplify the thought process for problem solving by using the conventional flow theory of Ben Franklin. The conventional idea is that the flow of current starts at the positive terminal of the battery (red wire) and proceeds around the loop, until it ends up at the battery's negative terminal (black wire). The reason that current flows is because the battery is providing an electric force to the circuit through chemical means, and a path for the current flow exists through the components that are in series in the loop of wiring connected between the battery terminals. Theoretically we know the electrons are jumping from atom to atom toward the positive battery terminal, but it is more helpful to us, for problem solving, to use the analogy of water flowing through a pipe when thinking about the process of electric current flow in a wire.

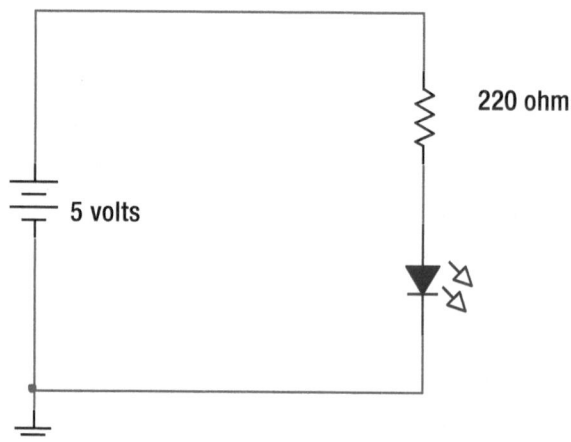

Figure 1-2. *An LED circuit*

The symbols used in our circuit drawing might look a little like ancient Egyptian hieroglyphics, but they actually make sense once we have a little background information. We call the symbols and diagrams *schematics*. The battery in the circuit is shown on the left, and the symbol is how a car battery looks inside, as seen from above with the water fill caps removed. The battery is comprised of a system of plates of metal surrounded by a sulfuric acid solution. One plate loses electrons, and the other gains them. This chemical action is responsible for setting up a positive charge on the one outside terminal of the battery, and a negative charge on the other. In our light-emitting diode (LED) circuit, the positive terminal is shown on the top. Because our circuit has a complete path of components and wiring, it is called a *closed* series circuit, which allows the battery's stored electric charge to flow as current through the components and wire around the loop. The first component that the conventional current flow encounters is shown as a squiggly line, which is the schematic symbol for a resistor. A resistor is used to restrict current flow. Again, the symbol is drawn to make sense, and can best be understood if you think of electric current flow being similar to water flow, and then thinking about how a stream or river zigzagging from side to side would tend to restrict, or *resist,* the free flow of water. The next component in the circuit that the current encounters is the LED, shown wired just underneath the resistor. It is a schematic symbol drawn as an arrow, because diodes only allow current flow in one direction. The positive and negative sides of the diode must connect toward their respective terminals of the battery or it will not light. The diode has *polarization,* whereas the resistor does not. The LED negative side can be identified as having the shorter lead, and also is represented by the side of the plastic component that is slightly flattened. Diodes work with electric current flow somewhat similarly to how valves work with water flow. In fact, in the very early days in the development of electronic diodes when they were vacuum tubes, they actually were called valves. Diodes have many purposes in electronics; when they are used to turn alternating current (AC) into direct current (DC) they are called rectifier diodes; diodes used to keep a voltage constant are called regulators or Zener diodes; and diodes used to oscillate at microwave frequencies and produce radio signals are called tunnel diodes. The LED is a diode that has the enhanced function to give off light as current flows through the polarized junction. The small unconnected arrows shown at an angle from the component in our schematic signify that it is an LED, with the arrows representing the light that is emitted from the device.

In the circuit, we have a 5-volt source, as this is the operating voltage of an Arduino Uno, and also the voltage it sends as a high level to its output ports (a port is a connection to the outside world). The unit of resistance is the Ohm, and resistors with higher Ohm values tend to restrict current flow more. The symbol for the Ohm is the horseshoe Ω, which actually is the uppercase Greek character Omega. So as to not overload the controller output, the value of 220 Ω will be used to limit current flow in many of our later Arduino projects. The resistor value does not need to be precise to illuminate a typical LED; one with a value in the 100 to 400 Ohm range will work fine. It's usually best to try to limit current flow as much as possible.

Using Ben Franklin's conventional current flow theory, the circuit operation is as follows: The positive charge on the high side of the battery terminal flows into the wire connected to the resistor. The resistor limits the current flow and drops the voltage. The LED that is connected between the resistor and the negative terminal of the battery lights with an intensity corresponding to the amount of current flow, as limited by the resistor. The LED will produce a voltage drop as well. When we talk about voltage drops, they occur across a component. When talking about voltage drops of more than one component, we add them together. Normally, the negative battery terminal is directly connected to ground, or the chassis, as it is in a vehicle. The symbol below the battery in our LED circuit represents earth ground. In residential house wiring, there is actually a long copper rod, 8 feet or longer, that is driven into the ground to establish the earth ground connection. Soil is somewhat conductive because of moisture and the salts and minerals it contains. Inserting the copper rod deeply into the earth provides much surface area contact with the soil and enables a good electrical connection. In automotive wiring, the same concept is used; however, in residential wiring the use of ground is primarily for safety concerns, whereas in a vehicle, the entire metal chassis is used as a return for the current to reach the negative terminal of the battery. The vehicle chassis is one half of the circuit path, so there is no need to run long lengths of wire to the negative terminal of the battery.

Now with all this background in electric current flow we can see why simplification is important to achieve a working knowledge of technology. Going back to the Bohr model of the atom and Figure 1-1 showing a copper atom, we know that the electron is charged negatively, and that when energy produced by a battery is connected to a closed circuit that current will flow. We were able to explain the operation of the LED circuit using Ben Franklin's conventional flow, and it makes sense because water runs downhill from a high level to one that is lower. Electrons actually flow uphill, though, because the negatively charged electrons are the carriers and move through the wires from a more negative, or low point, to a higher positive point where there is a deficiency of electrons. Thinking about water flowing uphill is hard to imagine, and the simplistic explanation of using conventional flow is incorrect, but it works and makes sense! A good analogy is that if you had one gallon of water per second flowing down a stream, or up a stream, either way you would have one gallon of water per second flowing in the stream. The numbers work out, and simplification keeps us from having nightmares about electrons jumping uphill, from atom to atom.

Ohm's Law

Just because a theory is old does not necessarily make it outdated or incorrect. The law we are about to look at was first published in 1827, and it remains in use to this day. Georg Ohm was a physicist who studied the relationship between the amounts of voltage, resistance, and current in electrical circuits. In science, there is a difference between a relationship and a law. A relationship signifies a linkage between values. A quantity might tend to increase or decrease as another quantity varies. If both values tend to rise together, then we can say that there is a *direct relationship*. If, however, one quantity increases as the other decreases, we would refer to the relationship as being *inverse*. In the last section, we mentioned that with a steady voltage, a larger value of resistance (measured in Ohms, Ω) would cause a decrease in current flow, and likened it to a zigzagging stream obstructing the path of water. The relationship between resistance and current is thus inverse. With constant voltage, you can look at this relationship in two ways:

1. As resistance goes up, current goes down.

2. As current goes down, resistance goes up.

The math symbol for proportionality resembles a fish (α). Usually in science, proportional relationships are found and then a constant of proportionality is used to make an equation that can then be solved for a numerical result. Just as in everyday life, relationships start out easy and get more complicated as stronger links are made. Luckily, Ohm's Law is not messy at all, and the relationships between voltage, resistance, and current turn right into equations without the need for a constant of proportionality. Ohm found the following simple equations to explain electricity (using V for volts, R for Ohms of resistance, and I for amps of current intensity):

$$I = \frac{V}{R}$$

$$R = \frac{V}{I}$$

$$V = IR$$

Figure 1-2, the schematic from the previous section, has been redrawn and is now shown as Figure 1-3, with the LED and resistor flip-flopped. The position of the components in a series circuit is irrelevant. The reason is because a series circuit is one loop, and all the current must pass through the wire and through every component in the path, regardless of the component's location within the loop. It makes the explanation a little clearer to design the circuit with the LED on top, because due to the internal construction of an LED, regardless of other circuit parameters, they tend to always drop approximately 2 volts. The reason there is a drop of voltage between the positive and negative sides of an LED is because a barrier junction is formed between the two sides that requires about 2 volts of force to push current through the device. The amount of voltage drop will vary with color; red has a little less drop and blue a little more, and the drop will increase slightly as current increases. Typical LEDs need about 20 milliamps (mA), which is two hundredths (0.020) of an amp of current to properly illuminate; some need a little less and some need a little more current to achieve proper brightness. In a circuit we will later build later, 120 Ω resistors are used. The nice thing about a hand grenade, nuclear war, and electronics design is that you do not have to be exact, just close. Now using Ohm's Law to calculate current in our circuit, first subtract the 2-volt drop across the LED internal junction, write the proper formula, and plug the numbers into a calculator.

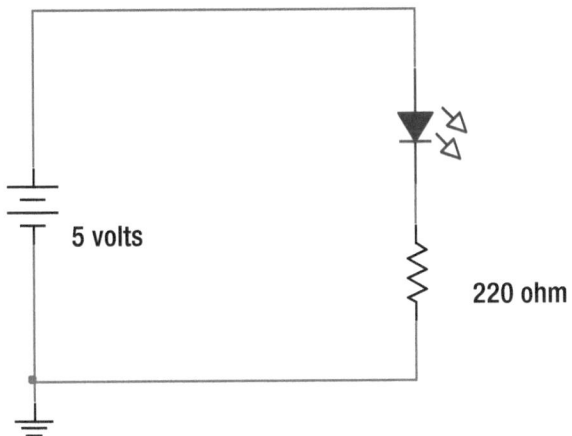

Figure 1-3. *Revised LED circuit*

5 – 2 = 3 volts across the resistor

$$I = \frac{V}{R}$$

$$I = \frac{3}{220}$$

$$I = .014 \text{ amps}$$

The current rounds off to about 14 mA, which is fourteen thousandths (0.014) of an amp. Because this is in series with the LED, it is also the LED current. Although this is only about three-quarters of the amount needed to bring the LED to full brightness, it will be visible. This will be a good Ohm value for our projects, as we will be connecting many LEDs to Arduino ports and need to keep currents to a minimum, so as not to overload the controller's maximum output.

Some people find it easy to have a graphical method to aid in finding the proper Ohm's Law formula to use with a given problem. The procedure for using the wheel shown in Figure 1-4 is to cover the unknown quantity, and the other two variables appear in the proper position to write the formula.

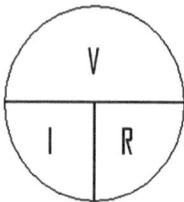

Figure 1-4. *Ohm's Law wheel*

It is interesting to note that if one were to graph a series of results with one independent variable held constant, and the other were to vary in the Ohm's Law formula for current:

$$I = \frac{V}{R}.$$

We find that with R held steady as V varies, a graph of a linear equation results because it is of the form $y = x$.

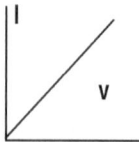

We also find that with V held steady as R varies, a graph of a hyperbola results because it is of the form $1/x$.

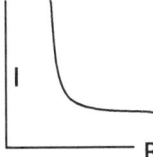

Along with the first quadrant graph, as shown for the linear equation, one could graph negative results located in the third quadrant, dependent on the frame of reference; however, there is not a true negative current, other than that as being referenced to its direction of flow. The hyperbola could also have a similar graph located in the fourth quadrant if the frame of reference of the fixed voltage was negative, which then caused current flow in a negatively referenced direction. Interestingly, both asymptotes of the hyperbola could never touch either on the axis because both continue to approach infinity, and the curve could never touch the origin, as there is no perfectly zero resistance.

Engineering Notation

In the last section, we said that LEDs typically require approximately two hundredths (0.020) of an amp of current for full brightness. Although the current requirement will vary greatly depending on the size, color, and lumens of output brightness, it will normally be in a range from 0.010 to 0.040 amps for common LEDs. Expressing quantities such as this in tenths, hundredths, or thousandths is very cumbersome, so in engineering the way of expressing large and small quantities is in a slightly different format than is used in scientific notation. To make things simple, engineering notation requires numbers to be in groups of three. Each group of three numbers is given a word prefix, so that they can easily be understood. When we discuss the large amounts of voltage and power the energy companies generate, we use two of the word prefixes, *kilo* and *mega,* attached to the units. For power, we have the names *kilowatts* for thousands of watts, and *megawatts* for millions of watts. The following is a list of engineering prefixes for both large and small numbers used frequently in electronics:

For large numbers:

$$Trillion = Tera = x\ 10^{12}$$

$$Billion = Giga = x\ 10^{9}$$

$$Million = Mega = x\ 10^{6}$$

$$Thousand = Kilo = x\ 10^{3}$$

The exponent $\times\ 10^{0}$ is assigned to the first group of three numbers, ending in 999, which are just units with no engineering prefix used. The following prefixes are for fractionally small numbers between 0.999 of a unit and 0.000000000001 of a unit:

$$Thousandth = milli = x\ 10^{-3}$$

$$Millionth = micro = x\ 10^{-6}$$

$$Billionth = nano = x\ 10^{-9}$$

$$Trillionth = pico = x\ 10^{-12}$$

In our LED circuit design problem in the last section (see Figure 1-3), we would say in engineering terms that the current in the circuit is calculated to be 14 milliamps (mA).

Review Questions

1. LED voltage drop will vary with color. (True/False)

2. One Meg-Ohm represents what value resistor in Ohms?

   a. 1,000 Ohms

   b. 10,000 Ohms

   c. 100,000 Ohms

   d. 1,000,000 Ohms

3. In Ohm's Law, resistance and current are:

   a. directly related.

   b. proportionally related.

   c. inversely related.

   d. the product of sums.

4. A diode that is used to turn AC voltage into DC voltage is called a _____ diode.

5. The unit of current is the _____ and the unit of power is the _____.

6. Conventional current flow goes around a closed path starting at the _____ terminal of a battery and ending at the _____ terminal.

7. Fifteen thousandths of an amp would be called what in engineering notation?

   a. 1.5 milliamps

   b. 15 milliamps

    c.   1.5 microamps

    d.   15 microamps

8. Explain the difference between science and technology.

9. The particle that carries current through a conductor is a(n)

    a.   electron.

    b.   mooseon.

    c.   proton.

    d.   nucleus.

10. A coefficient turns a mathematical relationship into a(n) _____ that can be solved.

Project 1

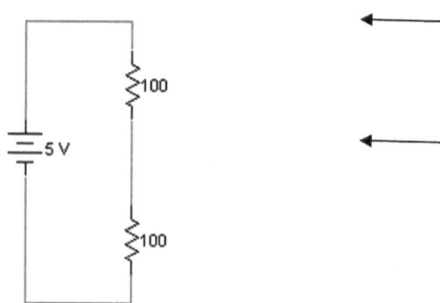

$Vr = 2.5$ is the voltage across the top resistor

***Figure 1-5.** A voltage divider*

Problem

Using Figure 1-5, find the current flowing through the wire.

Solution

One solution is that because it is given that 2.5 volts is dropped across the top resistor in the circuit, you can find the current flowing through it by using Ohm's Law. (You have V and R, so solve for current I.) Because this is a series circuit, the current is the same everywhere in the loop so the current through the top resistor will be of equal value to the current flowing through the bottom resistor, and also through the wire.

(Answers to the review questions and problems can be found in the Appendix of this book.)

CHAPTER 2

Computers and the Binary System

Digital Signals

The abacus could be thought of as the first computing device. It was developed in China more than two millennia ago and is used in some remote areas of the world to this day. Mechanical computing devices that worked on an analog basis eventually followed. The computing devices we use today are digital. The difference between analog and digital is shown in Figure 2-1. Analog signal voltage could slowly change with time, as displayed on the vertical y axis, with reference to time on the horizontal x axis, and analog signals could have a curved or a ramped wave shape, whereas digital signals rapidly jump between two discrete voltage levels. One of the digital levels is considered to be a *low* (0) and the other a *high* (1).

© Bob Dukish 2018
B. Dukish, *Coding the Arduino*, https://doi.org/10.1007/978-1-4842-3510-2_2

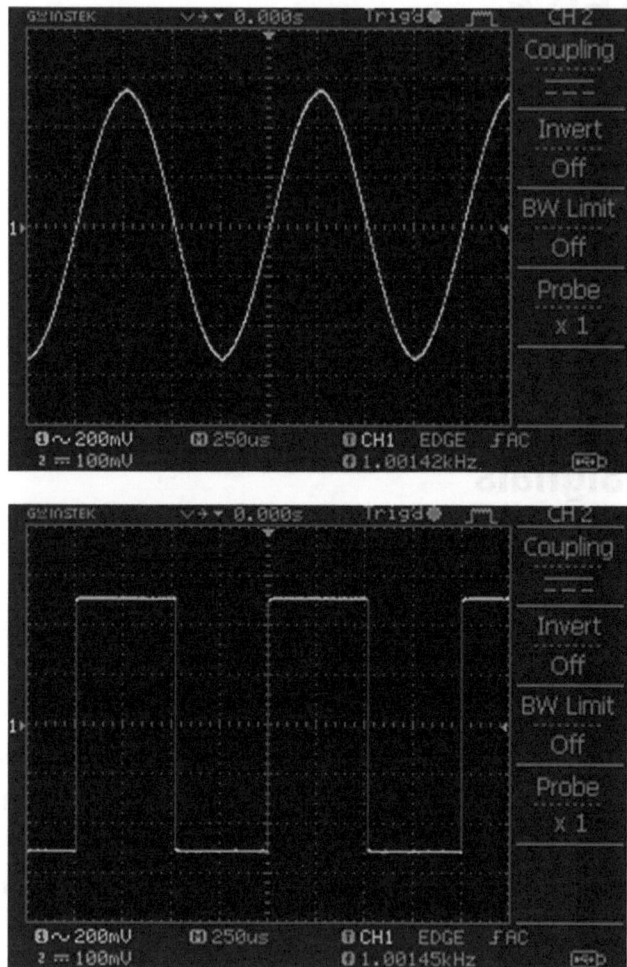

Figure 2-1. *Analog and digital oscilloscope displays*

The oscilloscope is a versatile testing device that allows the user to measure the voltages and time period, and to actually see one or more signals in an electronic circuit. By examining the bottom readings on the display, we see that the analog signal displayed in Figure 2-1 is approximately 1 volt from the top peak as measured to the bottom peak, which is called the peak-to-peak voltage. Voltage in a circuit is also sometimes referred to as *amplitude*. The frequency of the signal is approximately 1,000 cycles per second, called Hertz, so using engineering notation we could say it is a 1 kiloHertz (kHz) signal. If the generator producing the signal we are displaying in Figure 2-1 were connected to a speaker, you would hear a constant single tone. A 1 kHz analog signal is a standard test tone for audio circuits. Music is made up of a varying and complex waveform of many frequencies. Along with music and sound being analog, so is just about everything else in the universe, including light, gravity, heat, motion, and so on. Even the signals from digital broadcasts consist of analog waves that carry the signal and must be adapted to convey digital information through a process called *modulation.* Even most Internet connections (other than fiber) are analog and use modems, a name that is a contraction of the two words *modulate* and *demodulate.* Because the transmission carrier is analog, purely digital signals must be converted, or modulated, to adapt the transmitting carrier to convey the digital data. On receipt, the signal containing the data must be converted back to purely digital logic levels, or demodulated. The process of modulation and demodulation can be complex, but it is necessary to convey digital information over an analog medium. Fiber optic cable is a glass "pipeline" that can carry digital light pulses directly without modulation, but they are usually modulated to increase efficiency, and to allow for multiple transmissions to occur through the same fiber cable.

As we mentioned earlier, digital signals are either on or off, and there is no in-between state, other than for a brief transition time when the discrete logic level either quickly rises or falls. If the sun coming up in the morning were digital, one moment it would be dark, and the next moment it would be light. So, this question arises: Because practically no digital circumstances can be found occurring in nature, why are modern computers digital? The answer is that digital circuits are easy to design, manufacture, and miniaturize. As a digital signal is either on or off, it is very similar to the light switch on a wall, and a switch is the simplest electronic circuit. Also, there is no interpretation of the lamp brightness, as there would be with a light dimmer, which is an analog device.

Through the use of transistors acting as switches, digital devices can
be made very small and densely packed into *integrated circuits* (ICs),
commonly called *chips*. The very early digital computers used vacuum
tubes and electromechanical relays to act as switches, and they were quite
massive. One of the first digital computers was developed by the British
during World War II to help break the Axis Powers' Enigma code. It was a
code that was developed in Germany that was thought to be unbreakable.
The British computer's name was Colossus, and it was colossal. The
similarly massive American version was named ENIAC, and it contained
more than 17,000 vacuum tubes and 1,500 relays, and it used enough
wattage to power a large neighborhood. War seems to bring out both
the worst and the best in societies. Even the Internet came about from
ARPANET, which was a communications network designed for military use
during the Cold War, a time when the Soviet Union and the United States
risked mutual destruction. It was a precarious time for civilization, as the
annihilation of all life on the planet was a distinct possibility, but the Cold
War also led to the space race and perhaps one of humanity's greatest
achievements to date, humans walking on the moon. It was these early
military projects that brought about the computer revolution we enjoy
today. Seemingly the instruments of war many times are turned into useful
products and innovations that aid humankind through extraordinary
advancements in technology that benefit society. It seems that the bad
makes the good better. This yin and yang process occurs throughout the
universe and has an inherent cruelty when living organisms are involved
with the bad aspects of nature. There might even be some sadness from
older people waxing nostalgic over some of the early computer equipment
like 486 machines, 56K telephone modems, and picture tube monitors
that have met a horrible end, being ripped apart, smashed, and recycled to
provide the raw materials for the next generation of technical devices. As
advancements in circuit miniaturization have packed a whopping number
of transistors, in the billions, into a modern computer's processor, artificial
intelligence is now in its early stages. Soon machines might actually be

able to learn, think, and function without human intervention. What was once only science fiction could soon become science fact!

As we previously discussed, basic digital signals have electronic simplicity as shown in Figure 2-2. Along with that simplicity, though, also comes the added benefit of noise immunity. In the standard digital electronics designs that use the construction process of transistor-transistor-logic (TTL), the off state is called a low and is given the binary number zero, whereas the on state is called a high and given the binary number one. The voltages need not be exact, but must be below 0.8 volts for a low and above 2.0 volts for a high. Noise immunity comes about because any voltage can fluctuate between 0 volts and up to 0.8 volts and still be interpreted as a low, and any voltage can fluctuate between 2.0 volts and up to 5.0 volts and still be interpreted as a high. When we talk about noise in the computer and electronics field, we are describing electromagnetic interference. A good example of electromagnetic interference is the distortion of buzzes and howls sometimes picked up on an AM radio while listening to a ballgame or talk show from a distant station.

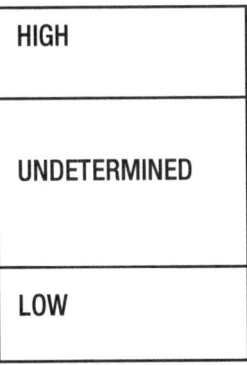

Figure 2-2. *Digital logic levels*

The oscilloscope is extremely helpful in analyzing analog waveforms, but a much simpler and very inexpensive piece of test equipment, called a logic probe, is extremely helpful in troubleshooting digital circuits when a technician is out in the field. It is just slightly larger than a pen and represents the two discrete logic levels by illuminating a red or green light. When analyzing multiple digital signals, a device called a logic analyzer is used. The logic analyzer is much like an oscilloscope, but able to display many simultaneous digital signals. The analyzer would be used in a shop setting, or during the design and manufacturing stages of digital products. It is helpful in troubleshooting timing skew issues, which sometimes occur in digital hardware when signals become out of synchronization due to unwanted delays, as the signals pass through electronic circuitry. Timing skew is a common problem during the design stage.

Power Consumption

The process of interfacing is necessary, as we already have discussed, when transferring information between digital and analog systems, but in the next two sections we are more concerned with interfacing voltages and currents and understanding the concept of power. In Chapter 1, you were introduced to the three variations of Ohm's Law, which mathematically explain the relationship among voltage, current, and resistance. The term *power* is also an electronics term but has two meanings when we talk about computers. One common definition is the amount of computational ability of a computer system. If in a conversational sense it is mentioned that a computer is a very powerful machine, the meaning that it is a high-end product that can process information, run programs, or operate very swiftly. In electronics, however, the term power has a very precise definition of a quantity. The watt is the unit of power, and it identifies the number of joules of energy consumed per second to produce work. Whenever we talk in engineering about units, we are referencing a basic quantity of measurement. In analyzing distances, we might use the units of inches, feet, and miles in the English system of measurement, with inches being the basic unit. The same concept holds true in electronics, where the Ohm's Law quantities that we spoke about earlier have the basic units of volts for the force of electricity, Ohms as the basic unit of resistance, and amps as the basic unit for current flow. This is a similar concept to using inches as the basic unit for distance. Just as inches can become feet, and feet can become yards and miles in distance measurements, seconds can become minutes, hours, and so on, but in science and technology the second is our standard reference for time. Again, power is energy consumed per second, and its quantity is given the name watt, named after James Watt, who invented the steam engine. If you are interested in exploring more of the field of electronics, you can look at the fundamental concepts that are usually laid out in great detail in physics books describing the nature of electricity and magnetism, and also from a more practical standpoint in books written for engineers and technicians. We might also recommend others in a fine series of books like *Extreme Fundamentals of Technology* and *Extreme Fundamentals of Energy,* both of which are written by the author of this text. For now, a short and quick explanation of power should suffice. There were three Ohm's Laws and there are also three power laws. Some people refer to them as Watt's Laws:

$$P = IV$$

$$P = I^2R$$

$$P = \frac{V^2}{R}$$

where *P* represents power in watts, *V* is voltage in volts, *R* is resistance in Ohms, and *I* is the current intensity in amps. Refer to Figure 2-3 for a detailed analysis using those quantities.

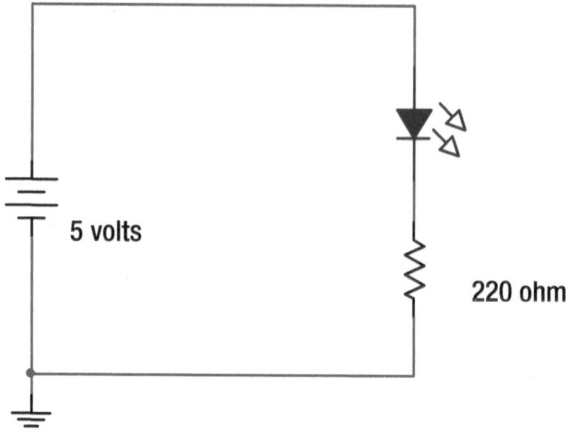

Figure 2-3. *Our LED circuit from Chapter 1*

We bring back our LED circuit that was used when exploring Ohm's Law, but now run power calculations. We can find the power consumed by the LED to produce light, as well as the power given off by the resistor as heat used to limit the current in the circuit. The process can be handled in different ways; one solution is to find the power used by each component and then add those amounts together. An analogy is that if you had two old-style incandescent light bulbs that were lighting a room in a house, and if each were a 100-watt bulb, the total power of the bulbs adds up to 200 watts. The same concept is used with LED bulbs, as well as other power-consuming devices. It is really a pretty straightforward concept. Using this method to find the total power consumption in our circuit, let's start by finding the power dissipated as heat in our current-limiting resistor.

You can use any of the three power formulas, but because we have learned that the LED drops about 2 volts, that means the resistor must be dropping 3 volts, because the total voltage of the battery source is 5 volts. Again, this is a pretty straightforward concept, but it is explained in more detail in other books and is known as *Kirchhoff's Voltage Law*. Finding the power dissipated by the resistor:

$$P = \frac{V^2}{R}$$

$$P = \frac{3^2}{220}$$

$$P = \frac{9}{220}$$

$P = .041$ watts, or we could say 41 milliwatts.

To find the power used by the LED, we can use Ohm's Law to solve for the current through the resistor, and because this is a series circuit, we know it is the same amount of current that is going through the LED. After we find the LED current and knowing its voltage is 2 volts, we can find its power consumption. We now find the current through the resistor:

$$I = \frac{V}{R}$$

$$I = \frac{3}{220}$$

$I = 0.014$ amps, or 14 milliamps (mA).
Now for the power of the LED:

$$P = IV,$$

$$P = (.014)(2) = 0.028 \text{ watts, or 28 milliwatts}$$

The total power is the power we found for the resistor plus the power of the LED. So, 41 milliwatts for the resistor + 28 milliwatts for the LED equals 69 milliwatts total. This is a very small amount of energy consumption, but many electronic circuits use small quantities such as this. Just as in lighting a house, the amount of circuits in a device have a power usage that is additive, so as the amount of circuits increases, so will the overall energy consumption. (Additional series circuits make up what are called a *parallel circuit.*) Typical PC power supplies could be as high as 500 watts. As was previously discussed, engineering notation can express both large and small quantities, and many times electronic circuits might only have powers in the milliwatt range (thousandths of a watt), and currents in the milliamp range (thousandths of an amp). On the other hand, we commonly use large resistors in the kiloOhm range (thousands of Ohms) to keep the current limited in circuits.

Interfacing

Microcontrollers like the Arduino are not directly capable of providing even moderately high currents or power. Microwaves, toasters, and coffee makers each require in the neighborhood of 1,000 watts to operate, which at the 120 volts used as household voltage in the United States equals the following current:

$P = IV$, divide both sides of the equation by V and you get

$$\frac{P}{V} = I.$$

By using the commutative law of mathematics, we get:

$$I = \frac{P}{V}.$$

For a typical microwave that uses 1,000 watts of power:

$$I = \frac{1000}{120}$$

$I = 8.3$ amps of current, but with the Arduino only capable of outputting 20 mA per pin, which is 0.020 amps, we have a problem. It looks like there is no way to run our kitchen appliances with a microcontroller. Actually, though, microcontrollers are now embedded in many kitchen appliances and operate them indirectly through the process of *interfacing,* through which the microcontroller sends voltage at a low current to an intermediary device that actually operates the appliance. The simplest intermediary device is an electromechanical relay. Relays have been used in automotive vehicles for many years and are referred to as *solenoids* when used for an application such as starting an engine.

The starter motor of a vehicle must provide a tremendous amount of torque to rotate the gasoline-powered engine up to the speed that is needed to begin the combustion process. In our earlier discussion of conductors and copper wire, we assumed that copper was a perfect conductor and had no resistance, when in fact nothing in the universe is perfectly good or perfectly bad, as there is no such observable thing in the universe that is perfect in every aspect. There always seem to be trade-offs. Copper wire does actually have a certain amount of resistance per foot, albeit very low. The trade-off is that we must shorten the length or increase the diameter for its resistance to decrease. All metal wire conductors behave this way and without performing any calculations it is worth noting the relationship.

$$R = \rho \frac{l}{A}$$

where R is the resistance of the wire in Ohms, ρ (the Greek letter Rho) is the amount of resistivity of the wire due to the material composition, l is the wire length, and A is the wire cross-sectional area. It can be seen, then, that as the wire increases in length its resistance goes up because of the direct relationship. On the other hand, as the wire increases in cross-sectional area by increasing the diameter, the resistance goes down, as shown in the formula, because the relationship is inverse between the cross-sectional area and the resistance. Rho is one of the constants of proportionality that we discussed earlier in the text that makes a relationship into a formula (i.e., no matter what material is being used as a conductor, the relationship is that as its length increases its resistance goes up, and conversely as its cross-sectional area increases its resistance goes down.)

In a vehicle starter motor circuit, as depicted in Figure 2-4, a very large amount of current must be carried through a conductor from the vehicle battery to the starter motor, so because the conductor has some resistance, we want to have a short distance of a large diameter wire, or there will be a drastic voltage drop and power loss across the length of wire. Because the key switch to the starter motor is located in the vehicle driver compartment, it would be impractical to run a very thick wire that is needed to increase the cross-sectional area up and down the steering column, and this also would increase our wire length. Instead, a thinner wire is used to carry a small amount of current that is used to cause a relay coil to energize by producing a magnetic field that pulls a high current contactor, making it close and thus making a remote high-current connection. The connection through the contactor provides a conductive path directly from the battery to the starter motor. The return path to the negative terminal of the battery is through the vehicle's chassis, which we consider to be ground.

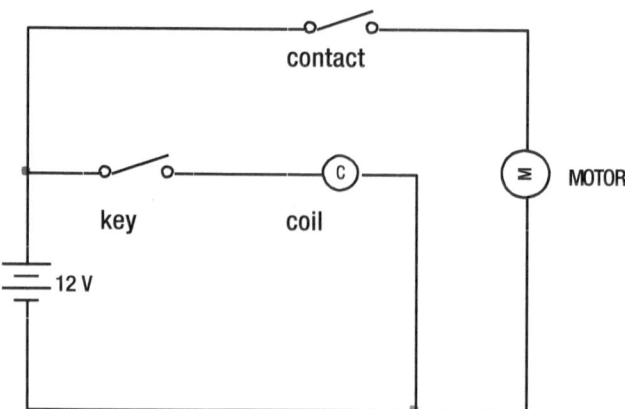

Figure 2-4. *Simplified vehicle ignition system*

The ignition system of a vehicle is shown as a simplified schematic diagram, where the positive terminal of a 12-volt battery is connected to both the ignition key and the starter motor contactor, with both shown in the diagram being in the off position, or what is called the *deenergized* state. When the driver turns the key, the key contact connection closes the coil circuit path and provides a current produced by the battery to flow through the relay coil circuit. The coil then produces a magnetic field that pulls the contactor in the starter motor circuit down and connects the path, which provides high current produced by the battery to flow through the starter motor. It then rotates and turns the vehicle engine, allowing combustion to occur, which will start the vehicle.

Relays are very useful as interface devices, but in most cases a small microcontroller like the Arduino does not even have the current-handling capability to energize a relay coil directly and might need a transitional interface circuit, called a *driver*. The term driver is also used in computers to mean a low-level software code needed to link a device with a computer operating system, or essentially a low-level software code needed to make some hardware operate. In electronics, we use the term to identify a device that will serve as an intermediary between two devices that cannot be directly connected together due to voltage, current, or resistance and impedance mismatches. In using the Arduino, or other microcontrollers, we might need to construct a circuit as shown in Figure 2-5 to electronically drive higher current loads than the microcontroller can accommodate directly.

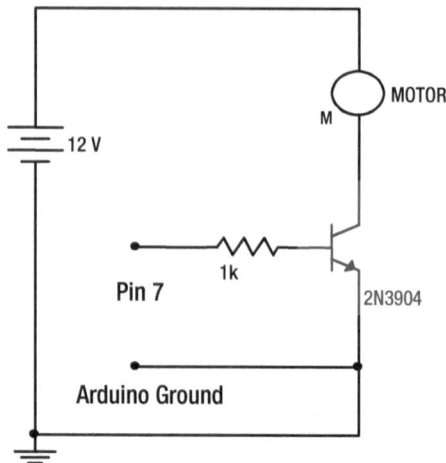

Figure 2-5. *Example of a microcontroller interfacing circuit*

This circuit could be used to run a very small 12-volt DC motor without a relay using the Arduino as a controller, as the voltages are kept separate with only the ground in common. In Figure 2-5, the left side of the open 1 kiloOhm resistor would connect to pin 7, or to any other digital pin that you wish to assign as an output on the Arduino. Because of the type of transistor that we are using, the Arduino would need to provide a logic high in order to run the motor. The control current being sourced from the Arduino can be calculated using Ohm's Law, and is found to be approximately 0.004 amps, which is 4 mA, well below the maximum recommended Arduino pin current of 20 mA. The value of 4 mA is found by knowing that the maximum voltage output from a logic high state on an Arduino UNO (the version we are using) is 5 volts. There is a voltage drop across the transistor of 0.7 volts from the middle pin to the bottom pin.

So, for our Arduino current path to ground, the voltage is 5 – 0.7 = 4.3 volts. Showing the calculations to find the current through the resistor:

$$I = \frac{V}{R}$$

$$I = \frac{4.3}{1000}$$

I = 4.3 × 10^{-3}, which is 0.0043, or 4 mA (rounding off). In electronics, it is very typical to round off numbers because there is usually quite a bit of component tolerance. The Arduino typically will output slightly under 5 volts, so the 4 mA result that was found is the worst case scenario of current draw. Also, notice in Figure 2-5 that the Arduino current path has no connection to the 12-volt source that we are using to run the motor, because that is a separate current loop. The control current acts similarly with the relay coil current that was previously discussed, in that it enables the top connection of the device to seemingly connect directly to the bottom connection. The bottom pin of the transistor, which is called the *emitter*, is attached directly to ground in our circuit. The middle pin that we have connected to the current-limiting resistor in the Arduino current loop is called the *base*. The top lead of the transistor connected to the motor is called the *collector*. The part number of the transistor is 2N3904. It is a small signal device that can handle a maximum collector to emitter current (the motor current loop in our circuit) of 0.2 amps, also referred to as 200 mA in engineering notation. If you wanted to drive a very large motor you could choose to use a transistor that can handle more current, or the intermediary circuit would need to be expanded to possibly include an electromagnetic relay, in which the transistor would have the relay coil in its collector to emitter path, and the motor current loop would connect through the relay's separate contactor. Most of the projects in this book do not require interfacing, as our main objective is to understand the programming thought process, but wherever you have a need to interface, knowing the three Ohm's Law formulas, which we covered, will help with your specific project.

Pull-Ups and Pull-Downs

The pull-up circuit shown in Figure 2-6 allows you to switch between a logic high and logic low, with the flip of a switch or the push of a button. The point shown to the far right in Figure 2-6 is the output and could connect to a digital input pin on a microcontroller. The value of the resistor need not be a specific value; anything will work fine in the range from 1 K, up to about 10 K Ohm or higher, depending on the IC. The output of the pull-up switch as shown in Figure 2-6 is called an *open*, and provides a logic high to the output. If the switch, or push button, were making contact (called *closed*), it would then connect to ground and give a logic low output. The resistor in the circuit is there to limit current flow so that there is not excessive current in the switch path when the switch enables a connection to ground (closes). If the switch and resistor were transposed, the circuit would be called a pull-down, and would normally give a low output, with the logic level becoming a high when the switch was engaged (closed).

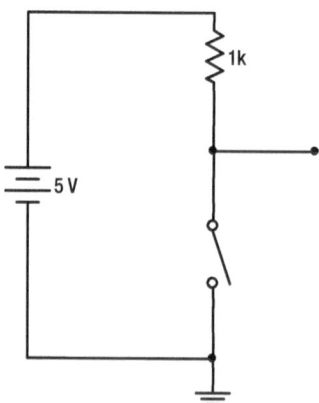

Figure 2-6. *Pull-up circuit*

When using mechanical switches to signal a logic level to a controller input, sometimes there is a problem with a phenomenon called *bounce* as represented in Figure 2-7. It occurs due to imperfections in the mechanical switching process in both pull-up and pull-down circuits. The contacts might make and break a few quick times in a split second before stabilizing. Normally a switch bouncing is not noticeable to us in the everyday world because it happens so quickly, but a processor can react at tremendous speed and might interpret bounce as more than one event. For example, if a program were meant to count each press of a momentary switch, the bounce issue might cause the processor to overreact and have an incorrect count that is too large. Most bounce problems occur within a time period that is in the very low millisecond range after a mechanical switching action has occurred. There are electronic hardware circuits that can eliminate the bounce problem, or you can compensate for the issue in your program code. Examples of coding methods to compensate for switch bounce can be as simple as adding a slight delay after a digital read command when a switch is to make or break contact.

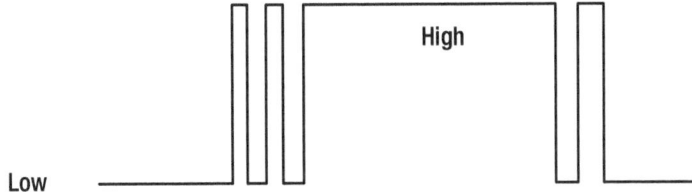

Figure 2-7. Switch bounce

Some sensors are analog, and the Arduino also has analog input/output (I/O) pins and excellent project examples in the integrated development environment (IDE). Because we try to use minimal hardware with this text, and because it is introductory, our projects involve mainly digital circuits, but we briefly cover analog projects later.

Review Questions

1. The TTL logic levels are a voltage below
   _____ for a low, and a voltage above
   _____ for a high.

2. Explain the difference between digital signals and
   analog signals, and give an example of each.

3. In the formula for resistance of a length or wire, the
   Greek letter ρ (Rho) for a given material is:

   a. the letter representing Ohms.

   b. the resistivity of the material.

   c. the length of wire.

   d. the power loss in the wire.

4. A pull-up or pull-down circuit prevents the
   excessive current of a short circuit. (True/False)

5. Explain what is meant by the term power.

6. A relay is an interfacing device that

   a. provides a direct path to ground.

   b. provides a remote connection.

   c. interrupts excessive current.

   d. does all of the above.

7. If you were to use a transistor as an interfacing device, the controller output would connect to which part of the transistor?

   a. the emitter

   b. the base

   c. the collector

   d. the gate

8. A 2N3904 transistor is very commonly used in interfacing the Arduino output pins to drive higher current devices. What is the recommended maximum current that the 2N3904 can drive?

   a. 200 mA

   b. 20 mA

   c. 200 amps

   d. 20 microAmps

9. Explain what is meant by the term noise in the computer and electronics field.

10. A very small and inexpensive piece of test equipment that could easily be taken out in the field to test logic levels of digital circuits is called

   a. an oscilloscope.

   b. a logic probe.

   c. a logic analyzer.

   d. a multimeter.

Project 2

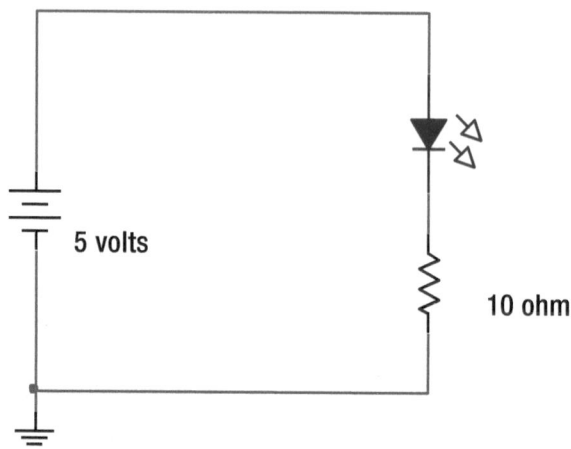

Figure 2-8. *An LED circuit with a small value resistor*

Find the current through the LED in the schematic of Figure 2-8. Is it an excessive current? (Hint, assume that the LED drops approximately 2 volts, which means that 3 volts is dropped across the resistor, as 2 volts + 3 volts = 5, and Kirchhoff's Voltage Law is satisfied. Also, remember that the current is the same everywhere in a series circuit.)

CHAPTER 3

Microcontrollers

Describing Microcontrollers

Let's start by discussing the differences between a typical computer and a microcontroller. Normal PCs have a mouse and keyboard for user input, and a monitor for output, whereas a microcontroller usually operates without human interaction. A typical computer could be used to multitask and run numerous programs simultaneously such as playing music, getting e-mail, and looking at informative web sites such as dukish.com, but the microcontroller can only run one program at a time (although the program can be redirected through the activation of hardware interrupts and software subroutines). Essentially a microcontroller is a specialized type of computer with mostly nonhuman I/O devices for interaction with the outside world. The job of a microcontroller might be to run the heating and air conditioning system in a building, monitor and control the operation of an automotive engine, or run machinery on an automated assembly line. Even though a microcontroller is not used like a personal computer, there are connections to the outside world. Rather than having a mouse, keyboard, and monitor as standard I/O peripherals, though, the microcontroller uses devices called sensors as inputs and actuators as outputs. A *sensor* examines the environment and senses specific stimuli, such as touch, sound, temperature, and other parameters, almost in the same way as human senses detect stimuli from the surrounding world. Microcontroller sensors respond to their respective stimuli by either creating a voltage, reporting a changing voltage, or generating an

© Bob Dukish 2018
B. Dukish, *Coding the Arduino*, https://doi.org/10.1007/978-1-4842-3510-2_3

abrupt change in a voltage logic level, and then sending the signal on a conductive wire to the microcontroller, as well as wirelessly through radio frequency (RF) links such as Wi-Fi and Bluetooth. In a human body, we have sensors to detect conditions in the outside world, which then similarly produce voltage outputs that are transmitted to the brain through the electrochemical wiring connections of the nervous system. The microcontroller can almost be thought of as a very tiny computer brain. Just as in the human body where the brain reacts to stimulus, the microcontroller sensors provide stimulus as input to a running program, which normally is running in a continuous loop, and the program can react and change outputs as conditions warrant. For instance, in the winter, if the temperature in a building goes lower than a preprogrammed value, a microprocessor-controlled thermostat will activate the heating system to increase the temperature to an acceptable level. This function, however, can be performed in a much more simplistic way without the necessity of a microcontroller, as was done for many years before the adoption of digital technology. Many older style wall-mounted thermostats contained a bimetallic strip, shaped into a circular spring pattern to control furnaces and air conditioning units. The two different metals on each side of the strip, each having a different temperature coefficient of expansion, caused the strip to bend and move the end of the metal spring to and fro in coordination with a changing temperature. The bimetallic strip spring was connected to a mercury switch, the movement of which either enabled or disabled the building heating and cooling devices by either making or breaking a connection used for control. In a microprocessor-controlled programmable thermostat, however, the set temperature can be altered due to factors such as the time of day or day of the week. Some of today's programmable thermostats even have network capability and can even be adjusted over the Internet as an *Internet of Things* device (IOT). The programmable thermostats are sometimes called "smart." It seems that, in many ways, the microcontroller operates in a somewhat human fashion. It is almost as though it is a one-track brain that can react, but not actually think. Due to their usefulness and low cost, microcontrollers have become commonplace in our homes, and are now found in such household appliances as home entertainment systems, refrigerators, and even high-end coffee makers.

Although computers and microcontrollers can mimic some human functions, please do not confuse their operation with true intelligence. We might call a thermostat "smart" or "intelligent," but all thermostats operate in a predestined manner only following a procedure as dictated by their programming. It does not matter how elaborate the device, or how good the program is that runs on the machine, computers and microcontrollers cannot truly be considered intelligent. Strides, however, are being made in machine learning and artificial intelligence. As multicore parallel processing, high operational speeds, and vast storage capacity in digital electronic devices are unfathomable to the generation that relayed on bimetallic strip technology to control heating and cooling in their homes, it seems that the next step might be a "thinking machine" that could actually become *sentient* (i.e., knowing that it exists). As for now, let's strive to understand how we can code a microcontroller to follow our commands and operate in a predestined and orderly manner.

The Arduino is an open source *prototyping* platform. A prototype is a first step in producing a new product, allowing for proof of concept testing that could lead to possible future refinement and production. The language that you will use to code the Arduino is a high-level computer programming language called *processing* or *wiring*. It is a slight adaptation of the very popular programming language C++, which has been around for quite some time, and is in widespread use in industry. C++ and its predecessor C are very powerful computer languages, and were even used to write sections of the Microsoft Windows operating systems. Microcontrollers in the past, such as the 8-bit Motorola (now Freescale) 65HC11, used a low-level programming method of operational codes (op codes), with specified addresses that were entered through the use of hexadecimal numbers, which are one step up from the binary one and zero machine language. This *assembly language* used a system of op code mnemonics instructions as a way to bridge the gap between humans and computers. Each processor had its own set of mnemonics and addresses, and it took a great deal of effort to write programs back in that era. Assembly is low-level programming because it deals very closely with the hardware and has direct control of the manipulation of data in digital circuits. Assembly language is a very powerful programming method that requires little in the way of speed and memory usage, and is surprisingly becoming quite popular again with the proliferation of IOT devices. Some examples of IOT devices are light bulbs, speakers, and even garage doors that can be connected through the Internet and controlled by computers and portable devices like smartphones. The high-level languages like C++, however, are much more common, and have both simplified the process of programming computers and microcontrollers and allowed for cross-platform compatibility and standardization.

The Arduino programs that you will construct with its open source freeware can be very complex, but we only need to explore a few of the features to get started writing useful code. We will be using the IDE for the Arduino. As we progress through this section, you will need to work on the IDE screen to write code and load it into the microcontroller, where it will continuously run whenever power is applied to the board. You will code in the white area under where it says sketch and the date. You code line by line to make the controller perform desired tasks. It is very important to understand that the instructions are executed in the line-by-line sequence of events. Once you are satisfied that your code should work, you can click the check mark button to discern if the formatting, or what is called the *syntax,* is correct. The syntax is very precisely laid out to make sure that the controller understands your exact intent. Once the code is in the proper computer language syntax, you can click the arrow located to the right of the check mark, or you could choose to use the menu bar, click File, and then click Upload to send your program to the Arduino board through a temporary USB connection.

Let's get a feel for how the IDE layout is structured (see Figure 3-1). The top horizontal section contains the menu bar, where you can open files, edit files, and so on. (Note that under the Tools menu, there will be a drop-down list for the *serial monitor* that will be used in the early stages of our program development in this text, but can also be used as an effective aid in debugging a program anytime you wish to check variable values, or know that a section of code has run as expected. The serial monitor can also be accessed by clicking the magnifier icon in the upper right of the IDE screen.) Located just below the words in the menu bar is the *toolbar* that provides shortcuts to the most common menu functions. This structure is very similar to many other Windows programs with which you are probably familiar.

Figure 3-1. *The IDE layout*

Very seldom will any program run perfectly on the first try. Even experienced programmers spend much of their time and effort testing and debugging program code. In any technological endeavor, troubleshooting is a normal procedure, and care must be taken to not become frustrated with the process. There are times however, after figuratively hitting the wall, that one might need to take a break and return later with a fresh perspective. It also helps developers to examine similar programs that are available in the public domain to see how others have approached solutions to difficult coding issues.

There are many sample programs located in the IDE under the File menu in the Examples section. The Blink program is a good place to start. If you have the IDE downloaded and installed on your PC and access to an Arduino board, you can practice uploading, running, and modifying the Blink program to change the blinking pattern. (The program is found under the File menu in the Example, Basic section.) In examining the Blink program code, notice that the command named `delay` has a number located between two parentheses. That number is the delay time in milliseconds. (Remember from our engineering notation section that *milli* is thousandths, so 1,000 milliseconds are equal to 1 second, because you have 1,000 thousandths.) You can easily modify the Blink program by either increasing or decreasing the delay number, which will lengthen or shorten the blink time pattern, respectively. You will also notice that the sentence statements end with the semicolon symbol rather than a period. This is true of Java and other popular modern languages and comes as a carry-over from an old programming language called Pascal. If you neglect to enter the semicolon symbol after each statement, or block of code, a syntax error will occur and the program will not run.

If you try to upload the Blink example code but it fails to operate, and a communications error occurs between the Arduino and the PC, you might need to change the *com port* setting in the hardware manager section of the PC operating system. This tends to be an issue that only happens when initially connecting the controller to the PC. There is a helpful IDE section that allows for changes in the com port. As shown in Figure 3-2, it can be found by selecting tools from the menu bar, and then usually selecting the highest numbered serial port that is listed.

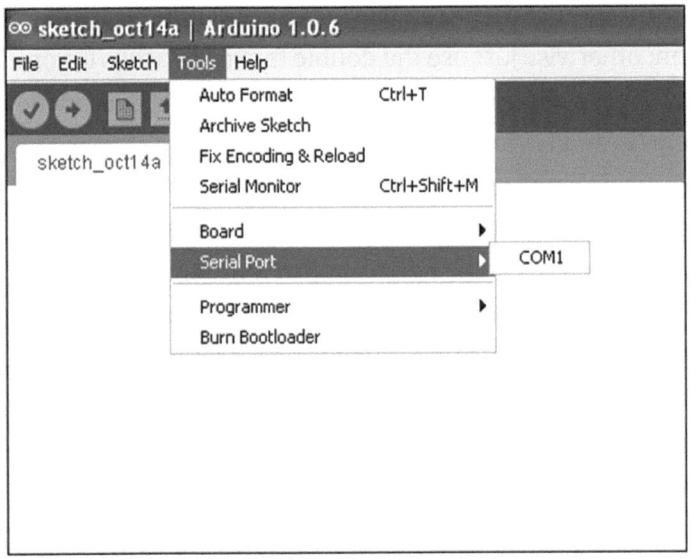

Figure 3-2. *Changing the com port*

Writing a Program

Most modern computer programming languages have similarities, and the material we cover with Arduino programming will fundamentally apply to other languages, with changes only in the syntax. The most important rule to remember in any programming language is to document your code. Whether you are working with a team of programmers, or you just want to make a few changes a year or two after you first wrote a program, providing documentation will help explain the intent of each section of the program. Most languages allow you to make comments in the code that are ignored by computers but can be seen and understood by programmers. In the Arduino language, we use double forward slashes (//) to begin a single comment line and pressing Enter on the keyboard ends the line. If you would like to generate a large amount of comments without beginning each line with the forward slashes, you can use a forward slash immediately followed by an asterisk (/*) to begin the comments, and switch them around at the end (*/).

Remember, this is only when you have a large section of document information; otherwise just use the double forward slashes for one line at a time. In the programs presented in this book, we might overdocument our code to help the reader, and there is no need for you to retype it into your test of the program. It is recommended that at this point you connect an Arduino UNO board to a USB cable linked to a PC running the IDE, and work along, as we present programs in the remaining sections of the book. The Arduino UNO can be purchased from many electronics outlets for approximately $25, or you can purchase it through the official Arduino web site at `http://www.arduino.cc` where the IDE is available as a free download.

After connecting the Arduino microcontroller to a PC with the IDE running, probably the trickiest thing is to connect to the proper com port. In Windows, the Found New Hardware wizard might open and ask to find the software. It is best to let it automatically search and click Yes to the questions. Afterward if there is a connection problem, as we just pointed out, it is a good trick to go to the Tools section on the menu bar, scroll down to Serial Port, and select the highest number port. This linking issue should only occur the first time you interface a new or different Arduino board to a PC. To check your link, use the menu bar to go to File, scroll down to Examples, then across to Basics, and select Blink. After examining the Blink code, click Upload next to the check mark icon on the IDE, and in a second or two, you will notice an LED located near pin 13 on the board blinking at the rate of about once per second. The UNO board has a built-in LED circuit containing a current-limiting resistor connected to pin 13. Having the ability to upload the program assures you have established a good connection.

In our first project, we will type code from scratch to control the blinking of the LED connected to pin 13. It is similar to the example code that is in the example Arduino software. To power the board, you can use USB power or disconnect the board after you have uploaded the program and connect a 9-volt battery to the *V in* pin on the board or to the power adapter jack. You can also connect 5 volts directly to the 5-volt header pin, but it must be precisely 5 volts or damage could result. Once the program is uploaded correctly, you will see a small LED located directly next to header pin 13 blinking slowly, slightly after the program has been uploaded. The uploaded program will stay in the board's memory and the LED will blink anytime power is applied to the board. To code our first program, begin by selecting File from the menu bar on the Arduino IDE program on the PC, and then select New. In the empty white code area, type in the code shown in Listing 3-1. To save time and effort, do not type the lines of documentation. In other words, don't bother entering any text between the characters /* and */ or anything on the single line following, and including the characters //. This is only documentation for yourself and other programmers that the processor ignores.

Listing 3-1. Coding the Blink Program

```
/*                    Blink the LED Five Times
This code blinks the LED connected to pin 13 five times and then
stops. It does this through the use of a while statement.
 */
// the declaration identifies variable "count" as interger, and sets
//size in memory
int count;
// next, the setup section assigns digital pin 13 as an output.
void setup() {
    pinMode (13, OUTPUT);
}
// the loop function would blink the LED forever if no while statement
void loop() {
  while (count < 5){ // goes for 5 blinks, zero through four
  digitalWrite (13, HIGH);   // turn the LED on with a HIGH
  delay(1000);              // wait for a 1000 ms (1 second)
  digitalWrite (13, LOW); // turn the LED off by making with LOW
  delay(1000);              // wait for a second
  count = count + 1; //adds one to the count of the number of blinks
  } //you can press the Arduino board reset button to start again

}
```

There are three sections that have been mentioned in the documentation. Examining each section and line of code, we notice the very top contains information about the variables used in the program. Remember that in algebra, a variable is a letter or group of letters that can represent a number. When writing a program, it is best not to use algebra letters *x, y, z,* and so forth as variables, but rather try to use letters that form a word that is descriptive of the variable's usage. This standard practice will help you understand the program's intent in using the variable. If you are working as part of a programming team, or you look back on your code in a year or two, it might be hard to remember what variables like the algebra letters *x, y,* or *z* meant, but is easy to realize that the variable `count` was used for a counter operation. All of the variables used in the program must be declared in the top section, however, and this is true in most programming languages. Also, the type of variable is declared, which allows the controller to allocate the proper amount of memory space for data storage, and also helps the processor know what type of operations can be performed on the variable. It was good past practice to set variables equal to zero when they were declared, but this is now unnecessary.

You might remember that in algebra there are constants along with variables. In science, a relationship might be found between two things, but to form an equation that can be solved for a numerical result, a constant of proportionality might be needed. We use constants in programming to help in the readability of code, just as we make variables descriptive. A constant could be used in the declaration section to refer to the LED pin, so that later in the code we would write LED, instead of pin 13. This is very helpful when there are many pins in use for inputs and outputs. In Listing 3-2, we modify our previous code by changing the number of blinks and speeding them up. We show the code without documentation, and the changes are highlighted.

Listing 3-2. Modifying the Blink Program

```
int count;
const int LED = 13; //assigning the variable "LED" to pin 13

void setup() {
    pinMode (LED, OUTPUT);
}

void loop() {
  while (count < 7){
  digitalWrite   (LED, HIGH);
  delay(500);
  digi talWrite (LED, LOW);
  delay(500);
  count = count + 1;
  }
}
```

The keyword `const` means the variable LED will always be associated with pin 13. If you left out the fact that it was a constant, the code would still run correctly because the Arduino is very forgiving, but it is good practice to include the `const` keyword. The keyword `int` assigns integer value and memory size. An integer is a number (with no decimal part) and includes zero. Two bytes of memory space are allocated, which allows the use of values from –32,768 to +32,767. If you are only using positive numbers (called whole numbers in mathematics), you can use the keywords `unsigned int`, which will allow you to use numbers between zero and 65,535. For the number 13 that we are using for the LED in our project, we could conserve memory space by using only one byte for the size, by using the keyword `byte` instead of `int`. A byte can hold numbers from 0 to 255. For numbers that have a decimal value the keyword `float` is used, which uses up four bytes of memory space. In modern PCs, memory size is not an issue, but in microcontrollers there is not much memory (only 2 K for the Arduino UNO), so we try to use memory conservatively.

Microcontrollers also run much slower than PCs, and have limitations on the number of internal bits that can simultaneously be processed. Typical PCs have quad cores, 64-bit data bus size (for simultaneous processing of the bits), and clock speeds of several gigahertz. The latest version of the Arduino uses an Intel Curie processor that has a dual core, 32-bit data size, and a 32 MHz clock speed. Originally designed for low power consumption useful in portable operation, it is an extremely efficient processor for the Arduino boards, but is out of production at this time. Microcontrollers generally do not need the fast processing speeds and functionalities of PCs because they run limited processes and have very efficient operating code.

In the second section of our modified code to blink the LED, we mainly assign inputs and outputs. The Arduino has bidirectional I/O. In our example, we want pin 13 to send out a voltage, and consequently a current through an onboard current-limiting resistor to the LED to illuminate it. In our tests, we are using the Arduino UNO board, which outputs the standard zero volts for a logic low, and 5 volts for a high. Some boards like the Arduino 101, which uses the Intel Curie processor, outputs 3.3 volts for a high. In either case, our code is the same and uses the keyword `pinMode`, followed by the pin number or name of the pin, and then the designation of input or output. All of the code we are using is case sensitive and special characters matter. If the words are misspelled or the formatting is off, the program will not run, but will instead alert you to a syntax error. If you typed the letter M lowercase in `pinMode`, or put a space between the two words, the code word not run. Syntax errors can be corrected by diligently reviewing every typed word, character, and punctuation mark on each line in the malfunctioning section of code. A logic error, on the other hand, is harder to catch because you have to rethink the overall approach to solving the problem.

The first two sections of code only run once, but the main loop will run continuously while power is applied. The main loop is considered to be a function, as is the set-up section, and the keyword `void` signifies that they perform an operation but are devoid of returning specific values after the process. The `while` statement is a conditional keyword that can invoke a section of code to run, such as in our program where a loop of code runs inside of the main loop. If the condition that the `while` function is based on is not true (or was true, but is no longer true), then the code located inside of the `while` function code space, which is identified by the nearest set of curly braces { and }, will not execute. In our program, the variable we called `count` starts at zero. Because code operates line by line, the `while` condition we specified is true as the variable called `count` starts at zero, so the LED turns on and stays on for 500 milliseconds before turning off and staying off for 500 milliseconds (500 thousandths of a second = 1/2 second). The count variable is then incremented so that the variable now goes from the value zero to one. The program loops at the back brace }, returning up to the `while` keyword, where the condition is checked again. The loops of on-and-off LED flashes therefore go through the counts of (0, 1, 2, 3, 4, 5, 6). When the counter hits the number 7, the condition in parentheses is no longer true (because 7 is not less than 7), and the `while` code space is exited at the line of code at the point of the back brace } for the `while` condition. Additionally, please note that the end of the main loop is the very last back brace }, and when the program hits that point, it is redirected back up to the point after the corresponding first main loop forward brace {. The `while` line runs on every main loop rotation, but because the counter is equal to 7 and it has not been incremented any higher or reset, the `while` keyword condition is false and the code inside of the `while` braces is skipped. The main loop keeps rotating, but not doing anything. It is completely normal to have trouble visualizing this detailed explanation of the code running line by line and it might be helpful to read this step-by-step description again slowly, and in small sections, while referring to the code listing that we are describing. The curly braces tend to be a main point of confusion when coding, but luckily the Arduino IDE can help you troubleshoot them. When you click just beyond a brace, the IDE will highlight what front brace it is associated with, or you can click just in front of the front brace to see its association with a back brace.

Computer processors cannot yet think for themselves, so programmers must tell them what to do every step of the way. It can be a little tedious and cumbersome, but after a little practice, it becomes lots of fun. This section of the book is also slightly tedious and cumbersome, but by taking time now and digging deeply into the foundation of how these things work, you will be able to use your creativity later to make powerful programs that can do amazing things. If, on the other hand, you find the background information too simplistic, hang in there because we are in a building process and are using scaffolding to get to the higher level material. Having a project that only flashes an LED five or seven times and then stops is not very exciting. Next, we will design code that will allow a push button to reset the blinking process.

There is a reset button on the Arduino that will restart the entire program, and in turn reset the counter variable back to zero, but there is a better way to restart the flashing without a hard reset. It would be impractical to reset the entire board if more code were running along with the flashing, so we will write some code that will use a push button to cause the program to reset. You might want to use a momentary switch and a breadboard if they are available, otherwise we can just use a 3-inch piece of wire connected to the input pin, and simply simulate a button push by momentarily touching the end of the wire to the metal shield around the USB connector located on the Arduino board. The shield is grounded and will give us a low logic level. We will write the code so that when a low is encountered, it causes an action to occur. To make the logic level change reliable, we invoke the *pull-up* mode in the setup section of the code. A pull-up can also be hardwired on a breadboard, but it is much easier to use the code to take care of this. The code in Listing 3-3 is similar to our last project but with the new material highlighted, so that if you are working along as you are reading the text, you will only need to modify the code by adding the highlighted sections. We try to follow this practice throughout the rest of the text, but sometimes we might greatly modify the code, so care must be taken that what is in the printed examples matches your code identically. Also, although not necessary, it is suggested that you use the same variable names as we do to reduce any confusion.

Listing 3-3. Further Modifications to the Blink Program

```
int count;
const int LED = 13;
const int button = 7;
boolean reset; //Boolean declarations take a small amount of memory

void setup() {
    pinMode (LED, OUTPUT);
    pinMode (button, INPUT_PULLUP);
}

void loop() {
  while (count < 7){
  digitalWrite  (LED, HIGH);
  delay(500);
  digitalWrite (LED, LOW);
  delay(500);
  count = count + 1;
  }
reset = digitalRead (button);
if (reset == LOW) {
count = 0;
}
}
```

This code causes the LED to go into the blinking process as soon as it is uploaded, or if the program has been uploaded and power is cycled off-on, or if the Arduino board reset button has been pressed. However, the new code that was added additionally allows for a programmed reset button to restart the LED flash sequence. It does this by resetting the counter to zero, so that the while condition is met and it causes that section of code space to loop seven times again (the counts are 0 through 6, which equals seven flashes). Notice in the top declaration section the variable `reset` was given the type `Boolean`. A Boolean number is only one bit, but is allocated one byte of memory in the Arduino. It can be used for our variable, as we are only dealing with a high or low logic level. Boolean logic consists of only two possibilities, 1 or 0, but you can also use the keywords `HIGH` or `LOW`. You can use the integer data type, but it is a waste of valuable memory space because integers are allocated two bytes. The variable could have been called *x* or anything you would like, but the variable name `reset` is descriptive. Some words, however, are reserved as keywords from the operational code and cannot be used. If you pick words that are reserved, the IDE program will show them in a different color, and will generate a syntax error causing the program not to upload or run. The logical reasoning for the resetting of the counter is as follows: Once the program has initially run, and then has stopped because the count has exceeded its max value to meet the `while` condition, the main loop continues looping while power is applied, so that when you momentarily touch the wire connected from pin 7 (which we called `button`), to the grounded USB box, it is read as a logic low. The keyword `if` is a very important conditional keyword! If a condition is true, then what is in the code space located between the open curly brace { and the close curly brace } after the `if` condition will execute; if the condition is not true, the code space will be skipped. When checking for a condition, this computer language uses the double equal signs (= =), but when doing math only a single equal sign is used. In our program, the controller will continue looping all day checking for this condition to be met. The situation is analogous to parents riding in the car on a long trip with their kids in the back seat periodically asking, "Are we there yet?" Anyone with a distorted sense of humor could modify this program to use a voice synthesizer to ask, over and over, "Are you going to push the button?" (I'll have to remember that for my next fun project.) It is also interesting to note that if you try to trigger a reset as the LED flash loop is executing, the program might ignore you if it is in the `while` loop and is busy performing the delay command. In the next modification, we interrupt the controller so that it is more reliable and responds immediately.

There are two types of interrupts: One is software defined and the other is activated by hardware. We will change the code to do the latter and create a hardware interrupt. On the Arduino UNO board, there are two pins that can be used as hardware interrupt inputs. They are pins 2 and 3, with pin 2 given the designation of interrupt 0, and pin 3 given the designation interrupt 1. We will move our wire from pin 7 from our last code, now over to pin 2, so that we will use interrupt 0. Interrupts do exactly what their name implies: No matter where the processor is in code execution, it stops to perform the interrupt service routine (ISR), and then returns to the point in the code where it stopped before the interrupt occurred. Also, note that if you are saving code as we work through our modifications, the IDE gives the program, also called a sketch, the current date as the default file name, but it is good practice to change the name to something more descriptive of your project. Also, the IDE saves work to the Arduino folder, but you might want to create project folders and save them in the Documents folder on your computer. Of course, it is an individual preference as to where and how to save files. Continuing now in modifying the original code from Listing 3-1, before we added the ability to reset the counter, we have the following new code (Listing 3-4) with additions highlighted for the use of a hardware interrupt. Momentarily tapping a wire from pin 2 to ground will cause a reset to occur immediately.

Listing 3-4. Using a Hardware Interrupt

```
//program to flash an LED seven times with interrupt reset
volatile int count; //volatile used for variables inside interrupts
const int LED = 13;
const int button = 2;

void setup() {
    pinMode (LED, OUTPUT);
    pinMode (button, INPUT_PULLUP);
    attachInterrupt (0, reset_ISR, LOW);
}
void loop() {
  while (count < 7){
  digitalWrite   (LED, HIGH);
  delay(500);
  digitalWrite (LED, LOW);
  delay(500);
  count = count + 1;
  }
}          ◄───────
// notice the ISR goes outside of the main loop
void reset_ISR() {
count = 0;
}
```

With the use of the interrupt, the program can be restarted at any point
during program execution. Interrupts are very useful when some event is
reported by an external sensor that demands an immediate reaction. In
our code, there might be a slight inconsistency in the number of flashes if
the program is reset during specific sections as it is running in the `while`
section of the code space. It seems that our program has encountered
a small bug. In the early days of the giant old mainframes that used
vacuum tubes and electromechanical relays, the warmth and light of the
tubes caused flying insects to be attracted to the computer equipment.
Occasionally a moth or other insect would get caught between relay
contacts or some other electromechanical device, and the computer would
need to be *debugged*. Nowadays our bugs are due to coding issues with
syntax or logic errors. We examine later how to disable interrupts from

occurring during critical sections of code and then how to reestablish them afterward. Our issue, however, is caused by the way that interrupts leave and then return to the main program. When an interrupt occurs, there is an internal memory called a stack pointer that remembers the exact point of program execution when the main program gets interrupted and exits to perform an ISR or subroutine. After the external code runs its course, the stack pointer shows the exact spot in the main program where the processor needs to return to resume its operation. The bug in our program deals with the *delay* function. If the main program was interrupted while it was into a delay, it will resume the main program at that exact point where it left the delay, which could be somewhere before or at the beginning, middle, or near the end. We need to digress to understand why we are even using the delay command in our program. When we give the command in our code to make an output pin a high or low the time period afterward is indefinite. The code in Listing 3-5 would only light the LED continuously. (Don't bother coding it.)

Listing 3-5. Lighting the LED Continuously

```
const int LED = 13;
void setup(){
pinMode (LED, OUTPUT);
}
void loop (){
  digitalWrite (LED, HIGH);  //the LED stays on from this point
}                            //as the program infinitely loops
```

Conversely, the code now shown will shut off the LED indefinitely:

```
const int LED = 13;
void setup(){
pinMode (LED, OUTPUT);
}
void loop (){
  digitalWrite (LED, LOW);  //the LED just stays off
}
```

Now, if we were to write the code shown in Listing 3-6 and upload it to the Arduino, we would notice a dim LED, because the on time and off time would occur very quickly, switching the LED on and off in a fraction of a second. Our human eyes would not discern the actual blinking light because of its rapid speed, but would instead interpret the overall intensity as half-brightness.

Listing 3-6. Changing the Switching Speed of the Blink Program

```
const int LED = 13;
void setup(){
pinMode (LED, OUTPUT);
}
void loop (){
  digitalWrite (LED, HIGH);
  digitalWrite (LED, LOW);
}
```

The delay command, as we have been using it to flash LEDs, can give a perceptible blinking effect. If the on and off times of the delay are somewhat short we can get a fast blink; if the on and off times of the delay are long we get a slower blinking of the LED. It would be a good learning experience to modify the code just presented to gain a good understanding of this process. In the next chapter, we debug our program glitch by using different types of delaying methods.

This chapter covered a large volume of material and we learned how to write code. It might be beneficial to stop here and review this section of text before proceeding. As mentioned in the book's preface, the best advice I ever received from a teacher was to reread complex material to gain a firm grasp of the content before moving on. In the next section, we will refine our programs, and have some fun communicating back and forth with a microcontroller. This might be a good spot to take a well-deserved break before we refine the code for microcontrollers to break out of loops and create delays that do not hold up the processor.

Review Questions

1. The main purpose of a microcontroller is

   a. to operate at faster clock speeds and have more memory than PCs.

   b. to test inputs and produce outputs.

   c. to run video games.

   d. to run simultaneous programs such as e-mail and Windows.

2. The Arduino is based on a version of the C++ language called processing. (True/False)

3. The Integrated Development Environment (IDE) is

   a. used to write programs.

   b. used to check syntax.

   c. used to upload compiled code to the controller.

   d. Used for all of the above.

4. Which are easier to debug: syntax or logic errors?

   a. Syntax

   b. Logic

   c. Both are equally easy to debug

5. A low-level programming method seeing an upsurge in popularity due to the proliferation of devices being produced for the Internet of Things (IOT) is

   a. Basic.

   b. C+.

   c. IDE.

   d. Assembly.

6. Misspelling or using improper grammar in programming code is

   a. a syntax error.

   b. a logic error.

   c. not using spell check.

   d. a loop.

7. The `while` conditional statement will run code contained between the forward brace { and the back brace } when the condition is not met. (True/False)

8. The main program loop

   a. continues looping anytime power is applied.

   b. loops one time.

   c. loops while the condition is true.

   d. loops while the condition is false.

9. The two types of interrupts are _____ and _____.

10. When an interrupt causes a program to leave its main loop where will it return?

    a.   at the beginning of the code

    b.   at the bottom of the code

    c.   at the next variable

    d.   at the next point where it left off

Project 3

Design a program that will flash the LED connected to pin 13 four times. Make the flash consist of 1 second on and 2 seconds off. Use pin 8 to reset the flashing so that it will restart when pin 8 is grounded.

CHAPTER 4

More Loops, and More Elegant Methods to Flash an LED

Timer Loops

The *delay* command is useful in most circumstances and it is easy to understand, but a `for` loop works very well as a timer whenever we want to repeatedly run a command a specific number of times, or when we wish to accomplish an additional task, or easily break out of the delay. Using a timer `for` loop should fix our delay bug from Chapter 3. In our program, our only additional task is to allow the possibility to break out of a delay, and leave the delay without returning there after an ISR or subroutine is activated. Note that both ISRs and subroutines are small subsets outside of the main program code. They both exist outside of the main loop, and are not used in all programs, but might sometimes be helpful. In the case of an ISR, the typically small section of external code responds to a hardware interrupt, such as that produced by a sensor or switch, whereas the traditional *subroutine* is mainly software defined and useful for organizing code so that repetitive procedures can be grouped and referred to periodically, rather than having repetitious code appearing numerous times within the main code loop. The use of traditional subroutines,

© Bob Dukish 2018
B. Dukish, *Coding the Arduino*, https://doi.org/10.1007/978-1-4842-3510-2_4

however, is not encouraged in C or C++. In our code for the program we are debugging (Listing 4-1), we use an ISR to reset the LED flash count. There are two possibilities for when this can occur: One is after the series of flashes have occurred and the LED is extinguished for a considerable period of time, and the other can occur as the flashing sequence is in progress. There are many ways to solve the issue of inconsistency of the number of flashes following a reset. One possible bug fix is shown in Listing 4-1. We are now executing a series of five slow flashes with the momentary tapping of a wire from pin 2 to ground causing a reset of the flash sequence.

Listing 4-1. A Possible Bug Fix

```
const int led = 13; //program for flashing an LED five times with reset
const int resetPin = 2;
volatile int reset;
volatile int loopCounter;
int counter1;
int counter2;
int insideLoop;

void setup(){
  pinMode (led, OUTPUT);
  pinMode (resetPin, INPUT_PULLUP);
  attachInterrupt (0, ISR_RESET, LOW);
}
void loop(){
while (loopCounter < 5){ //loop counter keeps track of total flashes
insideLoop = 1; //insideloop tells ISR that flashing was in progress

  for (counter1 = 0; counter1 < 100; counter1++){
    if (reset == 1){          //ISR makes reset 1
      break;                  //breaks out of the LED on for loop
    }
    digitalWrite (led, HIGH); //lights LED 1 second (10 ms x 100)
    delay (10);
  }
```

```
  for (counter2 = 0; counter2 < 100; counter2++){

   digitalWrite (led, LOW); // this section shuts off LED 1 second
   delay (10);
  }
  loopCounter++;      //add 1 to loopcounter
  insideLoop = 0;     //tells IRQ flashing not in progress
  reset = 0;          //allows LED on loop to run next time
 }
 }    //end of main loop
void ISR_RESET(){
  if (insideLoop == 1){
  loopCounter = - 1; //because on first exit after ISR one is added
  reset = 1;
  }
  else {
 loopCounter = 0; //starts the five flashes
  }
 }
```

The code uses the hardware interrupt as before, but now there are two loops to control the duration of each LED state, both contained within an outside loop used to control the total number of on and off flashes. We call the counter for the LED time on counter1 and the counter for the LED time off counter2. Each one repeats 100 times, with a delay of 10 ms, giving a 1,000 ms (1-second) total time duration, so that the LED is on for 1 second and then off for 1 second. For reset purposes, we use the variable called insideLoop to identify if a flash sequence is occurring. When there is a hardware interrupt, the ISR checks for this with an if else statement to restart the illumination sequence in the proper way. If the interrupt occurs during a flash sequence, the ISR returns with reset = 1 and the if statement that is embedded in the for loop of the LED-on will cause a break out of the loop, which cancels the LED on state. Because of that cancellation, the proper number of flashes will occur. The ISR had to restart the outside loopcounter variable at negative one to enable the proper number of flashes. Otherwise, if the interrupt occurs when the flash sequence is not running, the ISR simply resets the outside loopcounter variable to zero.

Notice that the on and off times were changed to 1 second each, and instead of seven flashes as in the program before, we changed it to five. We are now asking you to modify the code to restore half-second timing and seven flashes. The task lurks as an end-of-chapter problem, but it would be best performed now while the code is fresh. A hint appears in the Appendix.

An even more elegant method of causing an LED to flash while performing other operations in a program is to use a built-in timer function in the AVR processor used by Arduino. From the time that power is first applied to the board, the Arduino has a timer that begins counting the elapsed time. The counting process will continue for well over a month before it resets. To use this function, you need only specify the range either being micros or millis. The micros range would be used if you were measuring very short times in millionths of a second, whereas the millis range is for longer time periods that are in the thousandths of a second. For seconds of time we use 1,000 micros. In the next section of code, we use the timer so that we do not need to tie up the controller during the LED flashing process with a delay. This will allow it to perform other operations in the program code. This is an example of how we can mimic multitasking in a microcontroller.

Early PC operating systems used a similar method called *preemptive multitasking*. It seemed as though the computer was performing separate tasks simultaneously, but really it was only sharing time between different applications as required. In the very early days of mainframe computing, a similar system existed called *polling,* in which a specific amount of time share was given to multiple user terminals, so that they would connect for a brief period of time. It similarly gave the illusion that multiple events could be processed simultaneously. The difference between preemptive multitasking and time sharing is that preemptive multitasking only allocated processing time as was needed by the additional event, whereas mainframe polling would give an equal time slice to every operating terminal. Nowadays, with multicore processors and parallel computing, we are truly able to separately process data streams simultaneously. Even though some of the newest microcontrollers have multicore processors, their job is to have a one-track mind, and even with interrupts, subroutines, and elegant coding, the microcontroller is meant to run in a continuous loop.

Using the millis function, the code shown illuminates the LED, and then will extinguish it for a period of about 1 second each way, based on the last on or off state. It is not shown in Listing 4-2, but additional code could be written above, between, or below our LED code. Through this process, the controller can perform multiple tasks.

Listing 4-2. The Blink Program Using the Arduino Internal Timer

```
//blink program using the Arduino internal timer
const int led = 13;
unsigned long oldTime; //long used to allocate memory to large numbers
unsigned long newTime; // unsigned means no negative values
unsigned int timeDelay;
boolean ledOn;

void setup(){
  pinMode (led, OUTPUT);
}

void loop(){
  newTime = millis();
  timeDelay = (newTime - oldTime);

  if(timeDelay > 1000 && ledOn == 0){ //Note: && is the AND function
    digitalWrite (led, HIGH);
    ledOn = 1;
    oldTime = newTime;
    timeDelay = 0;
  }

  if (timeDelay > 1000 && ledOn == 1){
    digitalWrite (led, LOW);
    ledOn = 0;
    oldTime = newTime;
    timeDelay = 0;
  }
}
```

Controlling Embedded Processes

In the previous section on timer loops, the very last program that we examined used the `millis` timer to repeatedly flash the LED on and off. It was mentioned that other code could be performed while the flashing sequence was occurring so that it gives the illusion of multitasking. In this section, we demonstrate how the additional embedded code operates, and then later we modify the LED section of the code so that five flashes will take place and allow for a reset of the operation.

Because the Arduino only has one onboard LED, we need to connect an outside LED to one of the other header pins. Any other output pin would suffice, but our code calls for pin 12 as a second output to the external LED, and any one of the ground pins. You can just tie the two components together, or wire them on a breadboard. As shown in Figure 4-1, we need to connect the current-limiting resistor. You might wish to build two external LED circuits on a breadboard using different color LEDs, using pins 13, 12, and a common ground.

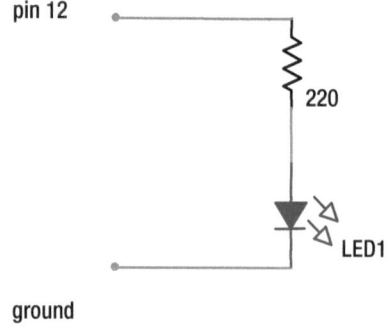

Figure 4-1. *Connecting the current-limiting resistor*

Listing 4-3. Using the Arduino Timer to Mimic Multitasking

```
//blink program using the Arduino internal timer to mimic multitasking
const int led = 13;
const int ledShort = 12;
unsigned long oldTime;
unsigned long oldTimeShort;
unsigned long newTime;
unsigned long newTimeShort;
unsigned int timeDelay;
unsigned int timeDelayShort;
boolean ledOn;
boolean ledOnShort;

void setup(){
  pinMode (led, OUTPUT);
  pinMode (ledShort, OUTPUT);
}

void loop(){
  newTime = millis();
  newTimeShort = millis();
  timeDelay = (newTime - oldTime);
  timeDelayShort = (newTimeShort - oldTimeShort);

  if (timeDelay > 1000 && ledOn == 0){   //one second LED on
    digitalWrite (led, HIGH);
    ledOn = 1;
    oldTime = newTime;
    timeDelay = 0;
  }
  if (timeDelayShort > 500 && ledOnShort == 0){ //half second LED on
    digitalWrite (ledShort, HIGH);
    ledOnShort = 1;
    oldTimeShort = newTimeShort;
    timeDelayShort = 0;
  }
  if (timeDelay > 1000 && ledOn == 1){   //one second LED off
    digitalWrite (led, LOW);
    ledOn = 0;
    oldTime = newTime;
    timeDelay = 0;
  }
```

```
if (timeDelayShort > 500 && ledOnShort == 1) { //half second LED off
   digitalWrite (ledShort, LOW);
   ledOnShort = 0;
   oldTimeShort = newTimeShort;
   timeDelayShort = 0;
 }
}
```

The program seemingly is controlling two simultaneous operations of two separate devices. To keep things simple, our second operation is just modifying the variable names and the blinking speed of the external LED, so that it blinks twice as fast as the internal LED connected to pin 13. Using this same concept with motors on a robotic arm instead of the LEDs, embedded code such as this could run two motors simultaneously, so that the arm could be made to run in a diagonal direction.

We now stop the process of the slower flashing LED, connected to pin 13, after five flashes with the following code, Listing 4-4, with changes shown as highlighted.

Listing 4-4. Blink Program Using the Arduino Internal Timer

```
//blink program using the Arduino internal timer
//two functions stops long flash on pin 13 at five flashes
const int led = 13; //long flashes on pin 13 onboard LED
const int ledShort = 12; //shorter flashing to external LED on pin 12
unsigned long oldTime;
unsigned long oldTimeShort;
unsigned long newTime;
unsigned long newTimeShort;
unsigned int timeDelay;
unsigned int timeDelayShort;
boolean ledOn;
boolean ledOnShort;
int longFlash;

void setup(){
  pinMode (led, OUTPUT);
  pinMode (ledShort, OUTPUT);
}
```

```
void loop(){
  newTime = millis();
  newTimeShort = millis();
  timeDelay = (newTime - oldTime);
  timeDelayShort = (newTimeShort - oldTimeShort);

 //one second LED on, for 5 total flashes on pin 13
 if ((timeDelay > 1000) && (ledOn == 0) && (longFlash < 5)){
    digitalWrite (led, HIGH);
    ledOn = 1;
    oldTime = newTime;
    timeDelay = 0;
 }
 if (timeDelayShort > 500 && ledOnShort == 0){ //half second LED on
    digitalWrite (ledShort, HIGH);
    ledOnShort = 1;
    oldTimeShort = newTimeShort;
    timeDelayShort = 0;
 }
 if (timeDelay > 1000 && ledOn == 1){ //one second LED off pin 13
    digitalWrite (led, LOW);
    longFlash = longFlash + 1; // keeps track of the long flashes
    ledOn = 0;
    oldTime = newTime;
    timeDelay = 0;
 }
 if (timeDelayShort > 500 && ledOnShort == 1){ //half second LED off
    digitalWrite (ledShort, LOW);
    ledOnShort = 0;
    oldTimeShort = newTimeShort;
    timeDelayShort = 0;
 }
}
```

The conditional statement that allows the slow flashing of the LED connected to pin 13 is expanded with the AND condition, so that the total number of flashes will not exceed five (i.e., the condition allows the LED to flash for the longFlash variable to increment five times: zero through the number four).

The next addition to the code will allow for a reset of the long flashing on pin 13, by zeroing the variable called `longFlash` through the use of an interrupt. The additional code is highlighted in Listing 4-5.

Listing 4-5. Using an ISR

```
//blink program using the Arduino internal timer
//two functions and stops long flash on pin 13 at five flashes
const int led = 13; //internal LED
const int ledShort = 12; //pin 12 used for external LED
unsigned long oldTime;
unsigned long oldTimeShort;
unsigned long newTime;
unsigned long newTimeShort;
unsigned int timeDelay;
unsigned int timeDelayShort;
boolean ledOn;
boolean ledOnShort;
volatile int longFlash; //ISR resettable variable
const int resetPin = 2;

void setup(){
  pinMode (led, OUTPUT);
  pinMode (ledShort, OUTPUT);
  pinMode (resetPin, INPUT_PULLUP);
  attachInterrupt (0, ISR_RESET, LOW);
}

void loop(){
  newTime = millis();
  newTimeShort = millis();
  timeDelay = (newTime - oldTime);
  timeDelayShort = (newTimeShort - oldTimeShort);

  if ((timeDelay > 1000) && (ledOn == 0) && (longFlash < 5)){

    digitalWrite (led, HIGH);
    ledOn = 1;
    oldTime = newTime;
    timeDelay = 0;
  }
```

```
if (timeDelayShort > 500 && ledOnShort == 0){ //half second LED on
  digitalWrite (ledShort, HIGH);
  ledOnShort = 1;
  oldTimeShort = newTimeShort;
  timeDelayShort = 0;
}
if (timeDelay > 1000 && ledOn == 1){   //one second LED off
  digitalWrite (led, LOW);
  longFlash = longFlash + 1;  // keeps track of the long flashes
  ledOn = 0;
  oldTime = newTime;
  timeDelay = 0;
}
if (timeDelayShort > 500 && ledOnShort == 1){ //half second LED off
  digitalWrite (ledShort, LOW);
  ledOnShort = 0;
  oldTimeShort = newTimeShort;
  timeDelayShort = 0;
}
} //End of main loop
//Interrupt Service Routines and Subroutines are outside main loop
void ISR_RESET(){
  longFlash = 0;
}
```

The code shown in Listing 4-5 adds a hardware ISR named ISR_
RESET, which will reset the counter called longFlash to zero. By
resetting the longFlash variable to zero, we allow the long flashing on
pin 13 to occur again for additional flashes. The interrupt could occur at
any time during program execution. The longFlash variable could be
reset at any time, including during the flashing sequence, to provide the
additional flashes.

The sections of the preceding code show examples of how seemingly
simultaneous operations can be coded, so that separate events can be
controlled in a microcontroller program. More than two events can also
be written into the program; however, as occurs with every processor, as

multitasking is expanded, the overall execution of the operations tends to slow down, which could cause issues in overall functionality.

Digital Electronics

If the coding programs we have presented to flash an LED on and off seem trivial by today's standards, then let's take a look at exactly how difficult an endeavor it is to accomplish through the use of basic electronic components. Please understand that the function we are examining could also be applied to many far-flung applications, with one such example being the control of vehicle intermittent windshield wipers. Until the late 1960s, the windshield wipers on vehicles had two speeds, low and high. Thanks to American inventor Robert Kearns, who patented a windshield wiper triggering system in 1964, whenever mist or light rain conditions developed that only needed a delayed sweep response of wipers across a windshield, an electronic circuit that he employed would delay operation for a preset period of time dependent on the values of two electronic components, the resistor and the capacitor. As we have learned in the first chapters of this book, the electronic component called a resistor is used in circuits to limit current flow. The other electronic component that Kearns used in his intermittent windshield wiper circuit, for timing, was a capacitor. A *capacitor* is used to store electric charge, and its charging function is determined by the amount of resistance in the series circuit path multiplied by the amount of capacity of the component. He developed a triggering circuit for windshield wipers that was delayed until enough charge built up across the components so that it would trigger a transistor to activate a wiper motor. In earlier chapters, we learned that transistors in computers operate as electronic switches, similar to wall switches used in our homes to switch lights on and off. The intermittent wiper is but one example to help illustrate that our ability to control the number of flashes of LEDs and their delay can be applied in solutions to numerous problems that exist in seemingly unrelated areas, and that the use of a preprogrammed microcontroller is one of many solutions.

A similar but far more robust transistor timing circuit than that used by Kearns in the 1960s was developed into an IC in the early 1970s. Its basic part number is 555, and it was used in early Apple and IBM computers. In fact, it is still widely in use today in many applications. Figure 4-2 shows one of its many uses, where it is wired as an asynchronous multivibrator. That function provides regularly timed pulses that continue indefinitely, so long as power is applied to the circuit. In our 555 schematic, we use values of the resistive and capacitive components that make an LED flash at roughly 1-second intervals (i.e., half-second on, half-second off). This is the same speed as our flashing LED in the previous programs. In later projects, we will speed up the frequency. The frequency of a 555 can be increased by decreasing the value of the resistive and capacitive components. The supply voltage does not affect the frequency. For a basic timing circuit, this hardware option is a low-cost solution when high accuracy is not a requirement.

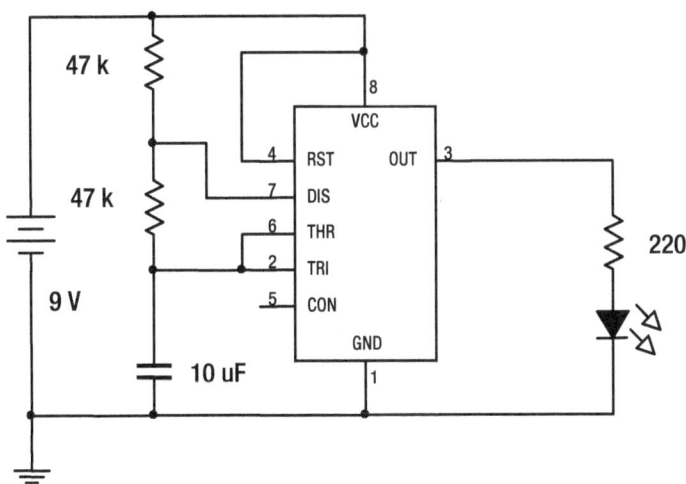

Figure 4-2. *Schematic of a 555 integrated circuit*

Typical part numbers for this IC timer are NE555 and LM555. The letters before and after the number identify the manufacturer and package type. The designation 555 is called the *base number,* and it is essentially the part number of the device. The timer is also available as a dual version with the base number of 556. Let's take a look at how the single 555 operates. We chose to use a 9-volt battery for our circuit because 9-volt batteries are readily available. The 555/556 can be operated at anywhere between 4.5 and 15 volts with an output current of up to 200 mA. Its wide range of voltages and currents, as well as its durability, is why this IC developed back in 1971 is still popular today. It is a hybrid of analog and digital technology and given the name *mixed signal.* If it were to be used with digital circuits the 9 volts would need to be varied accordingly (i.e., the voltage would need to be no higher than 5 volts for TTL logic devices, and for our Arduino). The schematic drawing is helpful to understand the operation of the 555 circuit, but the physical pins on the IC are numbered as shown in Figure 4-3. For all ICs in this type of package, which is called a dual inline package (DIP), the same pin numbering scheme is used, regardless of the total number of pins. Pin number 1 is always the leftmost bottom pin under the orientation marking, with the pin numbers increasing as they run, left to right, along the bottom of the package. The pin numbering then loops upward, counterclockwise to the top, and then runs right to left across the top of the IC package (see Figure 4-3).

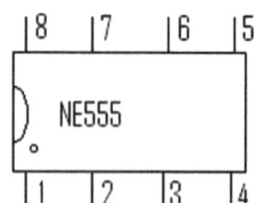

Figure 4-3. The DIP pin numbering scheme

In the 555 circuit of Figure 4-2, the main voltage (VCC) is connected from the battery to the power pin 8, as well as reset pin 4. Keeping pin 4 high does not allow the reset function to occur, and the 555 keeps flashing the LED from its output pin 3. In our example, the resistor-capacitor (RC) timing circuit consists of the two 47,000 (47 KΩ) series resistors, and the 10 microFarad (μF) capacitor (the Greek symbol μ, called mu, stands for the prefix *micro* in engineering notation). When power is first applied, charge across the capacitor is restricted by the two resistors that are in the series path to the capacitor. However, once enough time elapses and the capacitor develops enough charge across it, the 555 pins 2 and 6 trigger the device into operation, causing the capacitor to discharge into pin 7, which has an internal transistor turned on to supply a ground. After the capacitor voltage has discharged sufficiently to ground, the transistor inside of the 555 shuts off, and the pin 7 discharge path is then open, which allows the charge process of the capacitor to repeat. Because of its internal circuitry, the output of the 555 timer, which is pin 3, will oscillate between the VCC and ground voltage levels as a digital square wave. If you are building this circuit, do not tear it down because it could be used again in Chapter 8 as a frequency source for a number of electronic projects. To use the Arduino to provide power and ground, and for more information about the LM555 and breadboard connections, skip ahead to Chapter 8 and review the description.

To implement the logic of our Arduino project that ran a fast blinking and a slower blinking LED simultaneously, we could use a second 555 timer with different timing components, or just a single IC by selecting a 556 dual timer. In the 555 circuit just presented, the flashing rate was a half-second on, followed by a half-second off. The circuit used two resistors that were 47 KΩ, as are identified by the colors yellow = 4, violet = 7, orange = 3 zeros, which describes the value 47,000 Ohms of resistance. The gold band on the resistor is the last color band of four. It is helpful for orientation in reading the color code, and its gold color represents a tolerance of plus or minus 5%. Resistors are not polarized.

The capacitor value is directly labeled on the side of the component and it is polarized, so that the side marked with the negative symbol must go toward ground. (Most capacitors at or above 1 μF of capacitance are polarized.) To create a second 555 circuit to additionally flash at the rate of twice per second, we would need to construct a second circuit similar to the one in our schematic, but replacing the resistors and capacitor with values so that the RC time constant was one half. If we then ran both circuits simultaneously, it would accomplish the same outcome as our previous Arduino code. Alternatively, a single 556 dual circuit could be used for the same task. If that were the only function needed in a project, and the parameters were not going to change, it might be beneficial to only use two 555s or one 556 instead of a microcontroller because the size, cost, power consumption, and other factors could make the use of a processor on a small project impractical.

In the preceding exercise in digital electronics we can see that the control of objects and operations can be done electronically without the use of microcontrollers. Many legacy devices were controlled through the utilization of discrete components and ICs until recently, but with the cost of microcontrollers having become extremely low, and their versatility and processing power becoming very high, the use of a microcontroller is a simple solution when more than a few functions need to be accomplished, or if the parameters might change over time. As in our coding example, the flashing rate could very easily be adjusted through slight changes in a line or two of code, whereas in a digital circuit using discrete components, physical changes would need to be made. Even still, digital electronics is what make up the hardware of computer systems, cell phones, televisions, and other equipment we enjoy today. The study of both analog and digital electronics is a very useful and rewarding endeavor. The microcontroller is but a piece of the overall objective, which is to use technology for practical purposes.

Intermittent Windshield Wiper Control with Arduino

Intermittent windshield wipers were developed in the 1960s by Robert Kearns, as discussed earlier. There is an excellent movie about his invention and subsequent patent fight with the "Big Three" automobile manufacturers called *A Flash of Genius.* The popular film was released by Universal Pictures in 2009 and documents Kearns's tenacity. It is a movie that is educational, entertaining, and uplifting. In his invention, as we discussed earlier, Kearns used an analog timing system relying on the charge and discharge function of capacitors and resistors. The charge and discharge voltage level of the capacitor triggered a transistor into conduction for motor activation. Later, with the proliferation of digital circuits, and today with microcontrollers, a program can easily be coded to control the intermittent response of a vehicle wiper system. The program code that is presented in Listing 4-6 simulates the wiper motor direction by using two LEDs designated as wipeRight on pin 8 and wipeLeft on pin 9. In our simulation, each LED is separately connected in series to an Arduino output pin and through a 220 Ohm resistor, similar to Figure 4-1. Our code is rudimentary simulation. For actually using a nonfeedback loop DC motor, limit switches could be employed to compensate for motor drift. Servo, or stepper motors, could also be used to achieve a higher degree of reliability.

Listing 4-6. An Intermittent Wiper Motor Control

```
/*The intermittent windshield wiper implemented by a
microprocessor program. The pause is 3 seconds for mist
and 1 second otherwise
*/
const int startStopPin = 2;     //pin names
const int changeSpeedPin = 3;
const int wipeRight = 8;
const int wipeLeft = 9;
int wiperLatch;     //variable names
int startStop;
boolean toggleSpeed;
```

```
int dwellSpeed;
int count;
int delayChange;
void setup() {
  pinMode (startStopPin, INPUT_PULLUP);
  pinMode (changeSpeedPin, INPUT_PULLUP);
  pinMode (wipeRight, OUTPUT);
  pinMode (wipeLeft, OUTPUT);
}
void loop() {
  startStop = digitalRead (startStopPin);
  delay(200);
  if (startStop == LOW) {
    wiperLatch = 1;
  }
  dwellSpeed = 100;
  while (wiperLatch == 1) { //wipers are on.
    for (count = 0; count < 100; count++) {
      digitalWrite (wipeRight, HIGH);
      delay (10);
      subroutine();
    }
    count = 0;
    digitalWrite (wipeRight, LOW);

    for (count = 0; count < 100; count++) {
      digitalWrite (wipeLeft, HIGH);
      delay(10);
      subroutine();
    }
    count = 0;
    digitalWrite (wipeLeft, LOW);
    for (count = 0; count < dwellSpeed; count++) { //pauses here
      delay(10);
      subroutine();
    }
    count = 0;
    if (delayChange == 1 && dwellSpeed == 100) {
      dwellSpeed = 300;                 //3 sec dwell for mist
    }
```

```
    else if (delayChange == 1 && dwellSpeed == 300) {
//controls intermittent delay
        dwellSpeed = 100; //1 sec dwell for light  rain, starts by default

    }
    delayChange = 0;
  } //end of while loop
} //end of main loop

void subroutine() {
  startStop = digitalRead (startStopPin);
  if (startStop == LOW) {
    wiperLatch = 0;
    dwellSpeed = 100;

  }
  toggleSpeed = digitalRead (changeSpeedPin);
  if (toggleSpeed == LOW) {
    delayChange = 1;
  }
}
```

Review Questions

1. Use of the delay function is simple and straightforward, but it has a drawback in that

   a. the delay function is difficult.

   b. the delay function idles the processor from performing any other tasks.

   c. it requires many lines of code.

   d. it reduces the number of variables.

2. The letters ISR stand for

   a. Internet source reference.

   b. into some resistance.

   c. integer standing relative.

   d. interrupt service routine.

3. For loops are very useful if you wish to

    a. repeat a section of code for a specific time period.

    b. infinitely loop.

    c. delay and have the entire delay run without interruption.

    d. do all of the above.

4. The term multitasking in a microcontroller means

    a. seemingly performing multiple tasks simultaneously.

    b. performing tasks individually.

    c. running multiple clocks.

    d. using multiple variables.

5. If statements in a program are examples of

    a. conditional decisions.

    b. unconditional decisions driven by hardware.

    c. subroutines.

    d. interrupt.

6. The line of code that would have $x = x + 1$ in many high-level languages, equal to the code in the C++ language of X++, means

    a. the program is over.

    b. decrement x.

    c. increment x.

    d. the program begins.

7. An ISR is written

    a. inside a subroutine.

    b. outside of the main code.

    c. in the declaration section.

    d. in the setup section.

8. The code `break` is used to

    a. debug a program in IDE mode.

    b. end a loop or other function.

    c. stop for coffee.

    d. do none of the above.

9. The base number of an old but reliable timer IC still in use today is

    a. LM555CN.

    b. NE556XP.

    c. 555.

    d. NE556.

10. A polarized capacitor means that

    a. it has a positive and negative connection.

    b. its north side must face south.

    c. it can only connect to resistors.

    d. it is usually less than a value of 1 Ω.

Project 4A

In the first program that we covered in this chapter, we improved on the interrupt handling of the flashing LED programs from Chapter 3. We eliminated the long delay functions and replaced them with `for` loops. Please modify the code to restore seven distinct flashes, and so that the LED is on for one half-second and off for one half-second. (Hint: The code can stay essentially the same but some numbers will need to be modified.)

Project 4B

Design a digital circuit using the 555 timer as shown that will additionally illuminate a second LED during the opposite logic level. It should work like the warning lights at a train track, so that the two LEDs alternate their flashing. (Hint: The 555 supplies a high in the circuit as shown in the text to light the LED; however, it also provides a low when it is in its other state.)

CHAPTER 5

Serial Communications

The Binary Number System and ASCII Code

A very useful and fun tool in working with the Arduino is the serial monitor. Using the USB PC connection and the serial monitor feature of the IDE makes it unbelievably easy to use the on-screen serial monitor for both troubleshooting and creating entertaining game programs with minimal electronic interfacing. The IDE serial monitor function is bidirectional so that you can both transmit and receive data and see it in real time on the PC screen.

We know that computers only understand 1s and 0s, so for us humans to interact with our computers for writing an e-mail or posting to social media, for example, how do the words we type on the computer keyboard convert to the 1s and 0s of the binary system that our computers understand? The trick is to use a code that has a sequence of 1s and 0s to represent the information. It has gone through a few changes since the basic idea was developed in the early 1960s. Today the standard method uses 8 bits and is called ASCII code. The letters ASCII stand for the American Standard Code for Information Interchange (pronounced "as-key"). In using 8 bits and having the rule in the binary system that each bit may only contain a 1 or 0, the smallest 8-bit number is 00000000, and the largest is 11111111. As with the decimal system, characters to the left carry a greater weight. Also, each column has a specific weight associated with its location. In the decimal system that we use in everyday life, an example dealing with dollars in a paycheck is that the number 200 is much

© Bob Dukish 2018
B. Dukish, *Coding the Arduino*, https://doi.org/10.1007/978-1-4842-3510-2_5

better than the number 002, because the 2 in the first case is in the third
column where the weight of its position is in the hundreds of dollars,
whereas in the second case the 2 is located in the first column where it is
associated with ones. If you had a check for $202 that you took to the bank
to cash, the teller might give you two $100 bills and two $1 bills to pay out
$202. The binary system works in pretty much the same way, except you
cannot go above 1 in any column and the column weights are different.
The concepts are described using Table 5-1 and Table 5-2.

Table 5-1. *The Decimal Representation for Decimal 202*

10,000,000	1,000,000	100,000	10,000	1000	100	10	1
0	0	0	0	0	2	0	2

Table 5-2. *The Binary Number Equal to Decimal Number 202*

128	64	32	16	8	4	2	1
1	1	0	0	1	0	1	0

In Table 5-1, we have 200 + 2 = 202, and in Table 5-2 we have
128 + 64 + 8 + 2 = 202. So, if the bank were cashing your check in binary
money the teller would give you 11001010 dollars in binary. It looks like
a lot of money, but it is completely equivalent to the $202 that you would
have in the decimal system.

In our first serial monitor example, we use a simple Arduino program
to generate the decimal equivalent of the binary system *powers of two*. In
examining Table 5-2, we go from number 1, which is base two raised to
the exponent 0, up to decimal equivalent number 128, which is base two
raised to the exponent 7. If you count each of the columns in the chart, you
will find that there are eight of them, starting with the zero power, so the
chart contains 8 bits. Eight bits of data is also called a *byte*. In our following
program, the code expands the chart of base two, up to the 15th power

(16 bits, which is 2 bytes). The highest column has the decimal equivalent number of 32,768. After typing the code shown in Listing 5-1 into the IDE and uploading, click the magnifying glass icon in the upper right of the screen to go to the serial monitor, or you can choose Serial Monitor from the Tools menu bar.

Listing 5-1. Program to Generate Powers of 2

```
unsigned int i;   //program to generate powers of 2, up to 32k
unsigned int x;   //long int would be needed if you want to go higher

 void setup(){              //setup code only runs once
   Serial.begin (9600);
   Serial.println ("generating powers of two"); // prints title
   Serial.println();
 }

 void loop(){
   for (i = 0; i < 16; i++){ //16 loops, starting with exponent
                            //zero and ending with exponent 15
     x =  pow(2, i);

     if (i > 1){      //there is no direct power function in C++
       x = x + 1;        //the Arduino command for power has a
                       //glitch when used repeatedly.
     }           //after two entries we correct for the problem by
               //adding a one to the result with x=x+1
     Serial.print (x);  //x++ could also be used to increment x
     delay (1000);
     Serial.print("   ");
   }
   Serial.println();
   Serial.println();
 }
```

It gets a little confusing converting between different number systems; unfortunately, we have to go through the process because the everyday world is decimal, but computers are binary and can only deal directly with 1s and 0s. Fortunately for us, calculators and computers can easily make the translations. If you would like to perform the operation from our program on a scientific calculator, you can find each result by entering the base number 2 and then usually using the *x to the y* function to enter the power of two. It might be a second function key on your calculator. Usually the key is labeled x^y. You would enter the number 2 as the base, shown as *x,* then press the x^y key and enter the number for *y*, which would be the power of two for the exponent. You might need to consult the instructions for your specific calculator.

For a computer to understand what you are typing on a keyboard, there is an IC located inside of the keyboard case that, through the use of a matrix, associates each keystroke with its corresponding ASCII code number by generating a *scan* code. The code is needed because even though there are some numbers on a keyboard, there are letters, too (both lowercase and uppercase), in addition to punctuation and special functions. The group of experts who standardized the associative code gave a distinctive 8-bit number to represent all of the information on the keyboard.

Simulating Artificial Intelligence

In our next serial communication project, we will type a few letters on the keyboard and view the display on the serial monitor to see how the keyboard IC and the computer convert our letters to scan code and then to ASCII. There is also a behind-the-scenes interrupt number that is also sent from the keyboard to the PC, which is called an *interrupt request* (IRQ). It works similarly to the concept we used in the project when we sent a hardware interrupt to the Arduino to cause a reaction during our LED flash program. However, now the interrupt occurs automatically without having to poke a piece of wire to ground an input pin. It occurs when the keyboard sends data to the computer CPU for processing. The IRQ signal interrupts the CPU, so that it gives attention to the keyboard data that are sent. To observe this activity, we want to open the Arduino IDE and upload the program in Listing 5-2.

Listing 5-2. Programming for an IRQ

```
int keys;
int ASCII;

void setup () {
Serial.begin (9600);
}

void loop () {
if (Serial.available () > 0) {
keys = Serial.read ();
Serial.println();           //skips a space
Serial.print ("For the letter ");
Serial.print (keys, ASCII); //prints the letter
Serial.print (" the code in binary is ");
Serial.println (keys, BIN); //prints the binary number
Serial.print ("the decimal number is   ");
Serial.println (keys, DEC); //prints the decimal number
//you may wish to add code to print the hex number in this area:
}
}
```

After you have typed the code, which appears in the large open white area as shown in Figure 5-1, you can click the check mark on the left side of the IDE screen to test the syntax, or select the arrow next to the check mark in the upper left side of the screen, to both check and upload the code to an Arduino board. In either case, if there is a syntax error the IDE will display error messages and might highlight the code section that might have a problem. Sometimes syntax errors are not in the exact section that is highlighted. If you find that the results printed on the output screen seem to run together, or the syntax continues to have an error, a very common problem in this type of program is typing the print line return incorrectly. The command `Serial.println` has a lowercase *l* and *n* attached to the print statement. Many times, people mistake it for a lowercase *i* and *n*. The code in this example is pretty straightforward, but in general you will find that troubleshooting is a challenging and rewarding part of any development project. Quite often, errors in coding are due to a misspelled word or a punctuation error. Once the code is successfully uploaded, we are able to run the program. As you type letters on the keyboard to communicate over the serial monitor, the letter information is temporarily stored in a memory area called a buffer as it is transmitted serially as a packet to the PC when the Enter key is pressed, or the Send button is clicked. The Arduino can hold 64 bytes in its buffer, (i.e., 64 ASCII characters), and any keyboard information beyond 64 characters will be ignored and lost. In this program, we only need to enter a few letters to interact with the serial monitor and observe how computers translate between different number systems. To run the program and view the serial monitor, in the upper right corner click the magnifying icon (Figure 5-1), or on the menu bar click Tools and select Serial Monitor, to open the screen shown in Figure 5-2.

Now, type the lowercase letter *a* into the top input bar on the left and press Enter on the keyboard, or click the Send button on the right side of the screen. The monitor should then display the letter and both the binary code number and the ASCII decimal value for the lowercase letter *a* in the large open white area. You might also wish to add code to display the

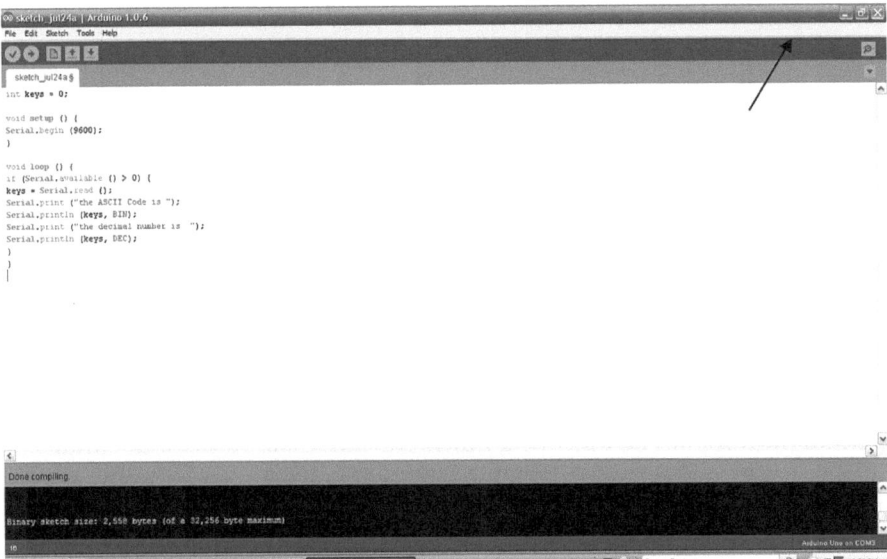

Figure 5-1. *Code entered in the IDE*

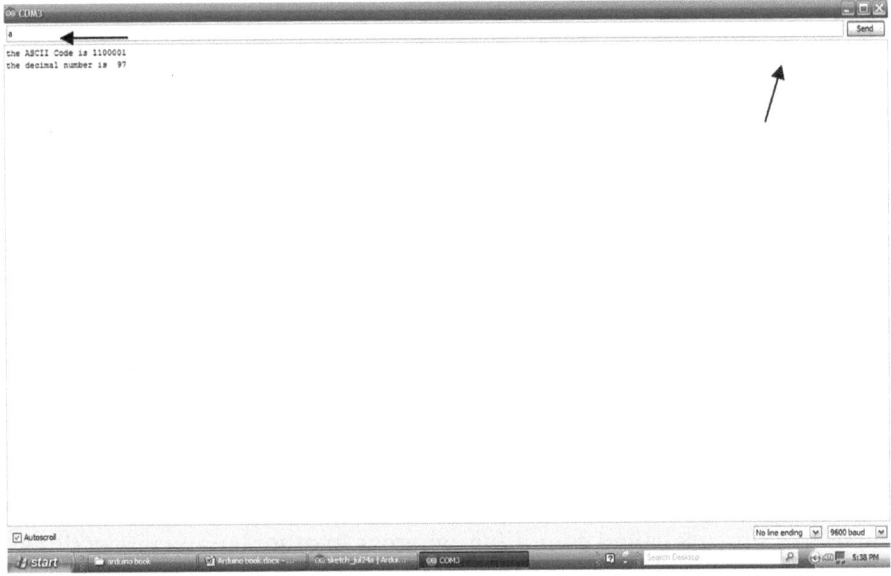

Figure 5-2. *The serial monitor screen*

hexadecimal (HEX) value. You would need to add a modified section of the decimal code from the program, where the DEC keyword is replaced by HEX:

```
Serial.print ("the hexadecimal number is");
Serial.println (keys, HEX);
```

The hexadecimal number system, usually referred to as just *hex,* is base 16, and uses 4 binary bits to represent a single number. Any binary number can be represented by a group of 4 bits in this format. Because 4 bits has its highest value equal to decimal 15 (1111_2), to only have one character, the hex system uses letters to represent single numbers above decimal number 9. The following numbers above the value of decimal number 9 are shown with their hex equivalent:

$$10 = A$$
$$11 = B$$
$$12 = C$$
$$13 = D$$
$$14 = E$$
$$15 = F$$

In mathematics, a base number system exists for every natural number (i.e., counting numbers), and the base is also referred to as the radix. In computers and electronics, we use the base 2 (binary system), base 8 (octal system), and base 16 (hex system), along with the normal base 10 (decimal system) systems. Computers only deal directly with the binary numbers, but octal and hex notation make it easier to write large groups of bits in an abbreviated manner. When we work with high-level programming languages, the conversion process is usually handled behind the scenes. In some languages, the entire program is converted into 1s and 0s in a process called *compiling,* whereas in other languages sections are converted only as needed, and that method is called an *interpreted* language.

When we input and output letters, printed text numbers, and punctuation directly with a computer program, each instance is called a *character* of the *string* data type. The decimal number of the ASCII code for some of the most common characters is shown in Table 5-3 and Table 5-4.

In our next serial monitor program example (Listing 5-3), we use the computer to both input and output information to the microcontroller. The program leads the user through prompts and outputs corresponding information about the user's and microcontroller's favorite colors. It is a very simple program that mimics artificial intelligence (AI). Work on

***Table 5-3.** Lowercase ASCII Letters*

a = 97	g = 103	m = 109	s = 115	y = 121
b = 98	h = 104	n = 110	t = 116	z = 122
c = 99	i = 105	o = 111	u = 117	Space=32
d = 100	j = 106	p = 112	v = 118	Period=46
e = 101	k = 107	q = 113	w = 119	Comma=44
f = 102	l = 108	r = 114	x = 120	Hyphen=45

***Table 5-4.** Uppercase ASCII Letters*

A = 65	G = 71	M = 77	S = 83	Y = 89
B = 66	H = 72	N = 78	T = 84	Z = 90
C = 67	I = 73	O = 79	U = 85	Numbers
D = 68	J = 74	P = 80	V = 86	zero through
E = 69	K = 75	Q = 81	W = 87	nine:
F = 70	L = 76	R = 82	X = 88	48–57

AI began in earnest after the 1940s when Warren McCulloch and Walter Pitts first mathematically quantified a *neural network,* the structure in which the neurons in animals' brains are wired together. A man-made neural network, or artificial neural network (ANN) is but one solution for creating a thinking machine. The Turing test for machine intelligence was developed at the beginning of the 1950s by the father of digital computer science, Alan Turing. The test consisted of a human having a conversation with either a computer or another human being, and not being able to distinguish any difference between the two. In 1956, John McCarthy, a professor at Dartmouth College, made the case that, "learning or any other feature of intelligence can in principle be so precisely described that a machine can be made to simulate it." That early era of computing gave way to children's robot toys and science fiction movies featuring thinking machines. Mathematicians, scientists, and programmers have feverishly been trying to develop AI in the years since then. Some forms of AI exist today, and as processing power continues to ramp up, intelligent machines will drastically change the landscape of our modern society. Today's best efforts to develop a thinking computer are to create a learning machine, in which a computer will learn from its past experience and pick the best future scenario. Seventeenth-century French mathematician Rene Descartes, well known for developing the Cartesian coordinate system, purposed the Latin philosophical concept, Cogito ergo sum, which roughly translated means, "I think, therefore I am." Today, as many well-known tech companies such as Google, Apple, Amazon, and even Facebook have been spending millions of dollars to develop AI, the single most promising field is *heuristics,* which is the ability to match patterns and relationships. There are also a great many computer programs that simply mimic machine intelligence. In our code, we add the ASCII decimal values of the letters in the words that the user enters on a keyboard. For example, the word *red* is comprised of the letters r = 114, e = 101, and d = 100. The total for the word *red* is thus 315. A glitch will occur in a more involved program using this simple method, due to the commutative property in arithmetic. Because our simple program does not take the sequence of the characters into account, a different series of numbers could happen to add together and produce the same total value. Listing 5-3 is merely a simple demonstration of the use of string I/O for multiple characters. Feel free to modify the code and make it more robust to more closely simulate AI.

Listing 5-3. Using String I/O for Multiple Characters

```
int keyboardByte;
int oldKeyboardByte;
int counter;

void setup(){
  Serial.begin (9600);
}
void loop(){            //displays an initial message
  if (counter == 0){ //can also put in the setup without a counter
    Serial.println ("");
    Serial.println ("I can respond to the word - hello typed and
entered above.");
    Serial.println (" ");
    Serial.println (" ");
    Serial.println (" ");
    counter = 1;
  }
  if (Serial.available() > 0){//keyboard keys pressed, data in buffer
    keyboardByte = Serial.read(); //reads a keyboard byte from buffer
    oldKeyboardByte = oldKeyboardByte + keyboardByte;
//stores letter as decimal value, and adds to total
  //Note:|| is the OR function. Found on most keyboards above "enter"
      if ((oldKeyboardByte == 532) || (oldKeyboardByte == 500)){
      Serial.println ("Hello there!");
      Serial.println (" ");
      Serial.println ("Are any of the colors red, green, or blue
your favorite color? (yes/no)");
      Serial.println (" ");
      oldKeyboardByte = 0;
    }
```

```
    if (oldKeyboardByte == 337){          //yes was selected
      Serial.println (" ");
      Serial.println ("Which of them is your favorite?");
      Serial.println (" ");
      oldKeyboardByte = 0;
    }
    if ((oldKeyboardByte == 432) || (oldKeyboardByte == 749)){
//user typed none, or none of them
      Serial.println (" ");
      Serial.println ("Sorry, I have to go");
      Serial.println (" ");
      oldKeyboardByte = 0;
    }
    if (oldKeyboardByte == 221){          //no was selected
      Serial.println (" ");
      Serial.println ("Oh Ok, well have a nice day");
      Serial.println (" ");
      Serial.println ("I only know the three colors");
      Serial.println (" ");
      oldKeyboardByte = 0;
    }
    switch (oldKeyboardByte){
    case 315:    //red
      Serial.println (" ");
      Serial.println ("Red is nice, but I like green");
      Serial.println (" ");
      Serial.println ("Nice talking to you, goodbye");
      Serial.println (" ");
      oldKeyboardByte = 0;
      break;
    case 529:    //green
      Serial.println (" ");
      Serial.println ("Green is my favorite color too.");
      Serial.println (" ");
      Serial.println ("Nice talking to you, goodbye");
      Serial.println (" ");
      oldKeyboardByte = 0;
      break;
    case 424:  //blue
      Serial.println (" ");
      Serial.println ("Blue is my second favorite color, but I
like Green better.");
```

```
    Serial.println (" ");
    Serial.println ("Nice talking to you, goodbye");
    Serial.println (" ");
    oldKeyboardByte = 0;
    break;
    }
  }
}
```

Text, called character strings, take lots of memory. If you are doing a project that is using lots of text printing to the serial monitor, you can try using a handy macro contained within the Arduino libraries, where the capital letter F encloses the text string; for example, `Serial.print (F ("This won't eat up RAM memory, since it goes to flash not directly to RAM") )`.

Designing a Serial Communications Game

In the last section, we examined the serial monitor, which is a very powerful tool for debugging code and watching real-time processing operations. The ASCII code described was more of a hardware process rather than the main objective of this text, which is to understand the software of programming controllers. The procedure was just presented as background information for the reader. There is low-level software called *firmware* that converts the keyboard scan codes and ASCII codes for us behind the scenes, so when programming a solution to a problem, we don't have to get bogged down and reinvent the wheel. If you type the number 5, as we learned, even though the information sent from your keyboard is only a code of 1s and 0s, that ASCII number represents the true number 5, so as high-level software programmers, we don't really care too much about the lower level code. The firmware will make all necessary translations. Computers operate at three distinct levels. The lowest level is the hardware level. The actual ICs and circuit boards of a computer are considered to be low level because they make up the machine level. The second level is the firmware and operating system level. This would be like Windows or Apple OS, where control of the hardware is maintained and interfaced with the top level. The top level is the user level, where human interaction occurs. As programmers, we produce programs that are at the top of this hierarchy.

We will now design a rather straightforward program that uses the serial monitor to give us output, thus saving us from having to interface electronic circuits to the microcontroller for the time being (Listing 5-4). We will design a game of over and under, where we will guess the outcome of a random number between 1 and 99 that the controller will generate. The code is provided in Listing 5-4 and an explanation follows.

Listing 5-4. Code for the Over and Under Game

```
/* Using a random number generator to get a number between 1 and 99,
   the player guesses if the next number is less or greater than
   the first number. It uses the serial monitor, and three digital
   I/O pins that are activated when momentarily touched to a ground
   point such as the USB shield box. */

int randomNum1;
int randomNum2;
const int button = 7;   //starts game round
const int lowButton = 8;   //player picks lower
const int highButton = 9; //player picks higher
int start;
int clocked;
int lower;
int higher;

void setup () {
  pinMode (button, INPUT_PULLUP);
  pinMode (lowButton, INPUT_PULLUP);
  pinMode (highButton, INPUT_PULLUP);
  randomSeed(analogRead(5));
  Serial.begin (9600);
  Serial.println ("Ground pin 7 to start ");//displays the message
}
```

```
void loop () {
  start = digitalRead(button);
  if (start == LOW) {                //game begins
    randomNum1 = random (1, 100);
    Serial.println (" ");
    Serial.println ("_____");
    Serial.println ("Playing between 1 and 99:");
    delay (1500);
    Serial.print ("The computer generated number is ");
    Serial.println (randomNum1);
    delay(2000);
    Serial.println (" ");
    Serial.println ("Do you think the next will be higher or lower?");
    delay(2000);
    Serial.print (" (momentarily ground Pin 8 for Lower, or 9 for
Higher) ");
    Serial.println (" ");
    Serial.println (" ");

    for (clocked = 0; clocked < 2000; clocked = clocked + 1) {
//waiting
        lower = digitalRead (lowButton);
        higher = digitalRead (highButton);
        if (lower == LOW || higher == LOW) {
          break;
        }
        delay (10);
    } //unbroken loop circulates 2000 times giving 20 seconds delay
    randomNum2 = random (1, 100);
    Serial.println (" ");
    Serial.print ("The Second number is ");
    Serial.println (randomNum2);
    Serial.println (" ");
    Serial.println ("*****");
    Serial.println (" ");

    if (clocked == 2000) {
      Serial.println ("SORRY TIME OUT"); //the player did not pick
    }
```

```
else if (randomNum1 == randomNum2) {
  Serial.println ("It's a Draw___Play Again");//randoms equal
}
else if (higher == LOW && randomNum1 < randomNum2) {
  Serial.println("You WIN !!! ");//picked higher and was higher
}
else if (lower == LOW && randomNum1 > randomNum2) {
  Serial.println("You WIN !!!"); //picked lower and was lower
}
else {
  Serial.println ("You Lose");
}
delay(2000);
Serial.println (" ");
Serial.println ("Ground pin 7 to play again ");
}        //end of game
}           //end of main loop
```

True random numbers are extremely difficult for a computer to generate.
A processor must follow a code leading to a predetermined sequence of
events. A trivial way a processor can pick a random number is for a group
of numbers to circulate in a loop. When the user pushes a button or causes
some sort of action, the number that is nearest the output of the circulating
loop is chosen as the pseudorandom number. It is not truly a random
number, however, because the user's choice of time determined the number
that was picked. There are many sophisticated pseudorandom generation
techniques. The best Arduino random number generator method is using
an analog voltage appearing at an unused analog input pin. It is simple
and works fairly well in generating random numbers. Electronic noise is
composed of stray signals that are out in open space like radio, TV, and
cell phone signals, or electromagnetic interference from electric lines and
the operation of household appliances, as well as signals produced by
solar wind and cosmic rays. The Arduino uses these random stray signals
to aid in the choice of a random number. In the setup section, the code
`randomSeed(analogRead(5));` reads the analog random noise value
of an open unused analog input pin and uses it as a reference to generate

a random number. You might be able to increase the reception of noise by connecting a small wire to the analog pin to act as an antenna. The `serial.begin (9600)` command opens the serial port on the Arduino and allows communication at 9,600 bits per second (the communications term bits per second is also called *baud*). We will monitor the activity just as we did when we used the serial monitor to examine the keyboard ASCII code. The variable named `button` (assigned to pin 7) starts the game when it momentarily goes low. A random number is then generated and user prompts are given, which appear on the IDE serial monitor screen.

Printing nicely to the monitor screen takes a little practice. The command `Serial.print ("X");` will print the letter X. Any string of characters enclosed in quotation marks will print directly on the screen. Without the quotation marks, the value of a variable *X* would be displayed as a decimal number. Adding `ln` (lowercase LN), shorthand for line, directly after a `print` statement will print the object, and then cause a line return by moving additional text to the following line. In the old days of teletypes, this was called a carriage return. To allow for better readability, you might wish to skip lines on the monitor output vertically, by using the commands `Serial.println ();` or `Serial.println (" ");`. An empty open and closed set of quotation marks is called a null string. It could also be useful to add horizontal spaces between text enclosed within quotation marks, such as when printing on the same line using the `Serial.print ("  ");` command.

The `for` loop in our code allows up to 20 seconds of time to elapse, for the player to make a decision, before the program times out. This amount of time is because each delay in the loop is 10 ms times the 2,000 times around the loop (.01 s × 2,000 = 20 s). The `break;` procedure will exit the loop as soon as a player decision is made so that we do not have to wait for an entire delay time to pass. We could have used interrupts, but the break works fine. If a condition is true, it breaks out of the loop and goes to the next section of code. The higher or lower pin choice is checked on each rotation of the loop. The `if` and `else if` conditional statements pick and display the proper outcome. Finally, a restart prompt is given, and the game can be played again.

Are computer games always fair? Let's modify the code to keep the player from winning by making a few minor changes. Add the changes highlighted in Listing 5-5 to our original program code and the player will never win.

Listing 5-5. Altering the Game Code

```
/* Using a random number generator to get a number between 1 and 99,
   the player guesses if the next number is less or greater than
   the first number. It uses the serial monitor, and three digital
   I/O pins that are activated when momentarily touched to a ground
   point such as the USB shield box.*/
int randomNum1;
int randomNum2;
const int button = 7;   //starts game round
const int lowButton = 8;   //player picks lower
const int highButton = 9; //player picks higher
int start;
int clocked;
int lower;
int higher;
void setup () {
  pinMode (button, INPUT_PULLUP);
  pinMode (lowButton, INPUT_PULLUP);
  pinMode (highButton, INPUT_PULLUP);
  randomSeed(analogRead(5));
  Serial.begin (9600);
  Serial.println ("Ground pin 7 to start ");//display message one time
}
void loop () {
  start = digitalRead(button);
  if (start == LOW) {                //game begins
    randomNum1 = random (1, 100);
    Serial.println (" ");
    Serial.println ("_____");
    Serial.println ("Playing between 1 and 99:");
    delay (1500);
    Serial.print ("The computer generated number is ");
    Serial.println (randomNum1);
```

```
    delay(2000);
    Serial.println (" ");
    Serial.println("Do you think the next will be higher or lower?");
    delay(2000);
    Serial.print (" (momentarily ground Pin 8 for Lower, or 9 for
Higher) ");
    Serial.println (" ");
    Serial.println (" ");

    for (clocked = 0; clocked < 2000; clocked = clocked + 1) {
//waiting
        lower = digitalRead (lowButton);
        higher = digitalRead (highButton);
        if (lower == LOW || higher == LOW) {
          break;
        }
        delay (10);
    } //unbroken loop circulates 2000 times giving 20 seconds delay
rerun:
        randomNum2 = random (1, 100);
        if (lower == LOW && randomNum2 < randomNum1) {
        goto rerun; //goes back and gets another second number
        }
        if (higher == LOW && randomNum2 > randomNum1){
        goto rerun; //goes back and gets another second number
        }
        //this section will not let you win
    Serial.println (" ");
    Serial.print("The Second number is ");
    Serial.println (randomNum2);
    Serial.println (" ");
    Serial.println ("*****");
    Serial.println (" ");
    if (clocked == 2000) {
       Serial.println ("SORRY TIME OUT");// the player did not pick
    }
    else if (randomNum1 == randomNum2) {
       Serial.println ("It's a Draw___Play Again"); //randoms equal
    }
```

```
else if (higher == LOW && randomNum1 < randomNum2) {
  Serial.println("You WIN!!!");//picked higher and was higher
}
else if (lower == LOW && randomNum1 > randomNum2) {
  Serial.println("You WIN!!!");//picked lower and was lower
}
else {
  Serial.println ("You Lose");
}
delay(2000);
Serial.println (" ");
Serial.println ("Ground pin 7 to play again ");
}        //end of game
}          //end of main loop
```

The highlighted code will reject the second computer-generated number if it is a winner and generate a new second random number. If the second random number would be a winner, the code uses a `goto` command to redirect the next line execution above the random number generation until a random number is picked that will cause the player to be wrong. Players will think their luck is very bad but should eventually catch on to the fact that they are being cheated, because they will never win a single round. In our next iteration of the game, we will occasionally allow them to win.

The changes to the original clean code first presented are again highlighted in Listing 5-6. Now we introduce a variable called `counter` to keep track of the number of times the second computer-generated number is repicked for the player to lose the round. We are introducing a complex type of division called *modulo* to allow the player to occasionally win a `round`. The modulo, sometimes called modulus or mod, is a division function that will only return the remainder of a division problem. So, if we mod divide a counter number by, say, 3, then only the number 3 and its multiples (6, 9, etc.) will have an integer result with no remainder. To make it visible for us to show where the cheating is happening, we are printing a message showing the number of times the second number was rerun to make the player lose (feel free to modify the code to change the odds).

Listing 5-6. Introducing the counter Variable

```
/* Using a random number generator to get a number between 1 and 99,
   the player guesses if the next number is less or greater than
   the first number. It uses the serial monitor, and three digital
   I/O pins that are activated when momentarily touched to a ground
   point such as the USB shield box. */


int randomNum1;
int randomNum2;
const int button = 7;   //starts game round
const int lowButton = 8;   //player picks lower
const int highButton = 9; //player picks higher
int start;
int clocked;
int lower;
int higher;
int counter;
void setup () {
  pinMode (button, INPUT_PULLUP);
  pinMode (lowButton, INPUT_PULLUP);
  pinMode (highButton, INPUT_PULLUP);
  randomSeed(analogRead(5));
  Serial.begin (9600);
  Serial.println ("Ground pin 7 to start ");//displays message once
}
void loop () {
  start = digitalRead(button);
  if (start == LOW) {                 //game begins
    randomNum1 = random (1, 100);
    Serial.println (" ");
    Serial.println ("_____");
    Serial.println ("Playing between 1 and 99:");
    delay (1500);
    Serial.print ("The computer generated number is ");
    Serial.println (randomNum1);
    delay(2000);
    Serial.println (" ");
    Serial.println("Do you think the next will be higher or lower?");
```

```
    delay(2000);
    Serial.print (" (momentarily ground Pin 8 for Lower, or 9 for
Higher) ");
    Serial.println (" ");
    Serial.println (" ");
    for (clocked = 0; clocked < 2000; clocked = clocked + 1) {
//waiting
       lower = digitalRead (lowButton);
       higher = digitalRead (highButton);
       if (lower == LOW || higher == LOW) {
         break;
       }
       delay (10);
    } //unbroken loop circulates 2000 times giving 20 seconds delay
rerun:
    randomNum2 = random (1, 100);
          counter = counter + 1; //starts count at 1 and 1 adds each time
          if(counter % 3 !=0){ //counter mod 3 not=0, cheat code runs
           if (lower == LOW && randomNum2 < randomNum1){
           Serial.print ("cheated   ");
           Serial.print (counter);
           Serial.println ("   times");
           goto rerun; //goes back and gets another second number
           }
           if (higher == LOW && randomNum2 > randomNum1){
           Serial.print ("cheated   ");
           Serial.print (counter);
           Serial.println ("   times");
           goto rerun; //goes back and gets another second number
            }
          }
    Serial.println (" ");
    Serial.print("The Second number is ");
    Serial.println (randomNum2);
    Serial.println (" ");
    Serial.println ("*****");
    Serial.println (" ");
    if (clocked == 2000) {
      Serial.println ("SORRY TIME OUT");// the player did not pick
    }
```

```
  else if (randomNum1 == randomNum2) {
    Serial.println ("It's a Draw___Play Again");//randoms equal
  }
  else if (higher == LOW && randomNum1 < randomNum2) {
    Serial.println("You WIN !!!"); //picked higher and was higher
  }
  else if (lower == LOW && randomNum1 > randomNum2) {
    Serial.println("You WIN !!!"); //picked lower and was lower
  }
  else {
    Serial.println ("You Lose");
  }
  delay(2000);
  Serial.println (" ");
  Serial.println ("Ground pin 7 to play again ");
  counter = 0;
  }        //end of game
}          //end of main loop
```

Finding Odd and Even Numbers

It was a bit of overkill to use mod divide in the last project, when other simpler conditional statements exist. We could have replaced the mod division with "while (counter < 3){...", or used "for (counter = 1; counter < 3; counter++)[{...". We wanted to introduce mod division, though, because it can be useful.

The next project finds even numbers using mod division by dividing an ever-incrementing number by 2 and checking for there to be no remainder (Listing 5-7). We use a wire quickly tapped to ground to increment the variable named value. When it is 2, or an even multiple, no remainder will be produced. The variable will print to the serial monitor screen as it increments, and pin 13 on the Arduino will light for 2 seconds if the number is even. Odd numbers would be a little tricky to find directly, but we could adapt the code to find them through the process of elimination (i.e., if the number is not even, then it is odd).

105

Listing 5-7. Finding Even Numbers Using Mod Division

```
int counter;
const int led = 13;
const int inputPin = 7;
int in;
int value;
void setup() {
  pinMode (led, OUTPUT);
  pinMode (inputPin, INPUT_PULLUP);
  Serial.begin(9600);
}
void loop() {
  in = digitalRead (inputPin);
  if (in == LOW) {
    delay(200);
    counter = counter + 1;
    value = value + 1;
    Serial.println (value);
  }
  if ((value % 2 == 0) && (counter == 1)) {
    digitalWrite (led, HIGH);
    delay (2000);
    digitalWrite (led, LOW);
  }
  counter = 0; //stops the LED from activating as the code loops
}
```

A Recipe Quantity Calculator for Baked Goods

We are again using the serial monitor, but now for a useful program.
The code is for baking recipes that allow the user to adjust for the exact
quantity of baked goods that are needed. The given code in Listing 5-8 only
has one recipe, but once additional recipe information is included in the
program, the specific recipe can be selected by looking for input equating
to the ASCII value of the names of the additional items.

Listing 5-8. A Baking Recipe Quantity Calculator

```
int keyboardByte;
int oldKeyboardByte;
int counter;
String number;
float total; //float allows for fractional values
int inputNumbers;
int muffins;
float x;
float flour;
float powder;
float soda;
float salt;
float butter;
float sugar;
float eggs;
float milk;
float vanilla;
float chocolateChips;

void setup() {
  Serial.begin (9600);
}
void loop() { //displays an initial message
  if (counter == 0){//can put it in the setup section without counter
      Serial.println ("Welcome to the Exact Baking Program");
      delay(2000);
      Serial.println(" (I only have muffins, and mcdonalds in my data
base, so far.)");
      delay(2000);
      Serial.println ("What would like to bake?");
      Serial.println (" ");
      counter = 1;
  }
```

```
  while(Serial.available()>0){//keys were pressed, data in buffer
    keyboardByte=Serial.read();//reads a keyboard byte from buffer
    oldKeyboardByte = oldKeyboardByte + keyboardByte;
//stores letter as decimal value, and adds to total
      inputNumbers = oldKeyboardByte;//reads a keyboard byte from buffer
      if (isDigit(inputNumbers)) {
        number += (char)inputNumbers;
        total = (number.toInt());
        oldKeyboardByte = 0;
      }
  }
  switch (oldKeyboardByte) {
    case 949: //mcdonalds
      Serial.println ("OK, mcdonalds it is!");
      Serial.println("If you're buying, I'll have two hamburgers and
large fries");
      Serial.println (" ");
      Serial.println (" ");
      oldKeyboardByte = 0;
      break;
    case 760: //muffins
      Serial.println ("OK, muffins it is!");
      Serial.println ("How many would you like?");
      Serial.println("(Please enter the number, followed by the
letter x) ");
      Serial.println (" ");
      oldKeyboardByte = 0;
      break;
    case 120: //x
      Serial.print("Here's the recipe for ");
      Serial.print(total);
      Serial.println(" muffins:");
      muffin();
      Serial.print (flour); Serial.println (" Cups of flower");
      Serial.print(powder);Serial.println("Tbsp baking powder");
      Serial.print(soda);Serial.println (" tsp of baking soda");
      Serial.print(salt); Serial.println (" tsp of salt");
      Serial.print(butter); Serial.println (" Cup of butter");
      Serial.print(sugar); Serial.println (" Cup of sugar");
      Serial.print(eggs); Serial.println (" eggs");
      Serial.print(milk); Serial.println (" Cup milk");
      Serial.print(vanilla);Serial.println("Tbsp vanilla extract");
```

```
      Serial.print(chocolateChips);Serial.println("Cup chocolate
  chips");
      Serial.println (" ");
      oldKeyboardByte = 0;
      number = "";
      total = 0;
      break;
    case 152: //space and x
      Serial.print("here's the recipe for ");
      Serial.print(total);
      Serial.println(" muffins");
      muffin();
      Serial.print (flour); Serial.println (" Cups of flower");
      Serial.print(powder);Serial.println ("Tbsp baking powder");
      Serial.print(soda);Serial.println (" tsp of baking soda");
      Serial.print(salt); Serial.println (" tsp of salt");
      Serial.print(butter); Serial.println (" Cup of butter");
      Serial.print(sugar); Serial.println (" Cup of sugar");
      Serial.print(eggs); Serial.println (" eggs");
      Serial.print(milk); Serial.println (" Cup milk");
      Serial.print(vanilla); Serial.println("Tbsp vanilla extract");
      Serial.print(chocolateChips);Serial.println("Cup chocolate
chips");
      Serial.println (" ");
      oldKeyboardByte = 0;
      number = "";
      total = 0;
      break;
      }
  } //end main
  //subroutine follows
  void muffin() {
      x = total / 12;
      flour = 2.5 * x;
      powder = 1 * x;
      soda = 1 * x;
      salt = .25 * x;
      butter = .25 * x;
      sugar = 1 * x;
      eggs = 2 * x;
      milk = 1 * x;
      vanilla = 1 * x;
      chocolateChips = 1 * x;
  }
```

Review Questions

1. If a `for` loop had a delay of 2, and a maximum loop count of 3,000, what would be the total delay in seconds?

2. Signals from unwanted electromagnetic waves are called

   a. noise.

   b. hyperbole.

   c. cosmic.

   d. digital.

3. The code `Serial.println ("Z");` will

   a. print the value of the variable Z.

   b. print the letter Z.

   c. print the value of the variable Z and then start a new line.

   d. print the letter Z and then start a new line.

4. `If` statements can be followed by `else if` conditions and may end with the last condition equal to `else`. (True/False)

5. The statement `myNumber = random (1, 10);`

   a. stores a random number between 1 and 9 into a variable.

   b. outputs the value of `myNumber` to the screen.

   c. picks the numbers 1 and 10 as random numbers.

   d. randomly loops between 1 and 10.

6. The base 16 system is also called

    a. radix.

    b. decimal.

    c. hex.

    d. 16 cycles.

7. An operating system such as Windows is which at level of the computer hierarchy?

    a. low level

    b. midlevel

    c. high level

    d. network

8. ASCII code has how many bits?

    a. 1

    b. 4

    c. 8

    d. 16

9. The hexadecimal number 1110 is represented by which letter?

    a. A

    b. F

    c. X

    d. E

10. The decimal number 9 is represented by which binary number?

    a.   1000

    b.   0009

    c.   1001

    d.   1111

Project 5

Modify the over and under game so that it is not possible for the second random number to equal the first. (Hint: Different methods are possible; one solution would be a `while` loop or a `do while` loop. The help section on the IDE contains a great reference section with examples of many different conditional statements.)

CHAPTER 6

Having Fun with Programming

Random Teacher Jokes

I would have to pay high taxes if this book project became a success, so in hopes of keeping sales down, it was decided to add a section making fun of teachers to ensure that the book would not be widely used in schools. The random number generator from the last chapter is a fun way to learn to develop games and interesting applications. In this section, we write code to deliver nasty random insults to teachers. You might wish to modify the code to poke fun of a lawyer, mother, or any other person of your choosing.

There are a few issues that we need to resolve in the program that follows. Some of the jokes do not make much sense because the insults in the first part do not match the punch lines in the second part. There is code to stop direct repeats of jokes, but the spacing is not far enough to eliminate repetition. One possible solution is to expand the joke library (please consult with a good comedian). Also, grouping related sections will tie the insults and punch lines together. Before we improve the code, let's understand how it works in its first phase.

Just as before, if you intend to copy and paste the following code from an e-book, be careful of the formatting of apostrophes and other punctuation marks. It might be best to just retype the code as shown into a blank Arduino IDE. Be very careful, however, because the IDE does

© Bob Dukish 2018
B. Dukish, *Coding the Arduino*, https://doi.org/10.1007/978-1-4842-3510-2_6

not provide a spell check or punctuation correction function. To help format the code after you enter it, you can press Ctrl+T on the keyboard. To run the program after it is uploaded to an Arduino, momentarily tap a wire from pin 7 to ground, such as the metal box surrounding the USB connector, so we can see an obnoxious, but possibly humorous joke materialize from code in Listing 6-1.

Listing 6-1. Teacher Joke Generator

```
byte randomNum1;        // Teacher joke program, first attempt
byte randomNum2;
boolean counter;
const int trigger = 7;
byte trigState;
byte trigLatch;
String firstPart;
String secondPart;
byte oldRandomNum1;
byte oldRandomNum2;

void setup(){
  pinMode (trigger, INPUT_PULLUP);
  Serial.begin (9600);
  Serial.println ("Tap a wire from pin 7 to ground for a joke");
  Serial.println ();
  Serial.println ();
  randomSeed (analogRead(5));
}

void loop(){
  trigState = digitalRead (trigger);
  if (trigState == LOW){
    trigLatch = 1;
    do{
    randomNum1 = random (0, 10);
    }
    while (oldRandomNum1 == randomNum1); // checks and eliminates
                                //redundancy in the first part
    do{
    randomNum2 = random (0, 10);
    }
```

```
  while (oldRandomNum2 == randomNum2); // eliminates redundancy
                                       //in the second part
  }
  oldRandomNum1 = randomNum1;
  oldRandomNum2 = randomNum2;

  while (trigLatch == 1){
    subRoutine1();
    subRoutine2();
    Serial.print ("My teachers are so ");
    Serial.print (firstPart);
    Serial.print (" they ");
    Serial.print (secondPart);
    Serial.println (" ");
    Serial.println (" ");
    trigLatch = 0;
    delay (500);
  }
} //end of main loop
void subRoutine1(){
  switch (randomNum1){
   case 9:
   firstPart = "fat,";
   break;
   case 8:
   firstPart = "big,";
   break;
   case 7:
   firstPart = "tall,";
   break;
   case 6:
   firstPart = "old,";
   break;
   case 5:
   firstPart = "cheap,";
   break;
   case 4:
   firstPart = "smelly,";
   break;
   case 3:
   firstPart = "ugly,";
   break;
```

```
   case 2:
   firstPart = "messed Up,";
   break;
   case 1:
   firstPart = "nasty,";
   break;
   case 0:
   firstPart = "skinny,";
   break;
   }
}
 void subRoutine2(){
   switch (randomNum2){
   case 9:
   secondPart = "block the sunshine";
   break;
   case 8:
   secondPart = "scare themselves";
   break;
   case 7:
   secondPart = "melt snow in the winter";
   break;
   case 6:
   secondPart = "make cabbage smell good";
   break;
   case 5:
   secondPart = "make water run up-hill";
   break;
   case 4:
   secondPart = "cause bill collectors to pay them money";
   break;
   case 3:
   secondPart = "make robbers run the other way";
   break;
   case 2:
   secondPart = "cause rocks to break apart";
   break;
   case 1:
   secondPart = "make quick sand slow";
   break;
```

```
  case 0:
  secondPart = "cause trees to lose their leaves";
  break;
 }
}
```

In analyzing the code, we again describe our variables in the declaration section, and in the setup section we define inputs and outputs and call for the random number generator. After the button is clicked to trigger the start of the program, we incorporate a `do-while` loop. We have used `while` loops in previous code examples, but the `do-while` is slightly different in that it performs an action and subsequently tests its condition. The thought process here is to store the random number from the previous run of the program and while the old number and new number are equal, the generator will run until they are not equal. In this way, we eliminate a direct repeat of both the insult and the punch line. Next, this program makes use of software subroutines that are frowned on by traditional C and C++ programmers. In early computer programming languages, such as Basic, subroutines were used for organization and to keep code modules from having to be repeatedly rewritten. Languages like Visual Basic, which is very intuitive and fun to use, rely heavily on subroutines to perform operations in conjunction with user inputs. Although now frowned on, subroutines work nicely for our objective by associating a *string* with a random number (strings are groups of characters that form words). In the subroutine sections, we use a common conditional statement that is similar to `if` and `else  if` called the *switch case*. `Case` statements are used in many languages to consolidate a large number of `if` statements. In our code, it matches the random numbers to the associated strings. We use the first set of random numbers to generate the insults and the second set of numbers to generate the punch lines as they are printed to the serial monitor. If you run the code in Listing 6-1, you can see that some of the jokes make no sense. In Listing 6-2, a grouping sequence is used so that there is more of a relationship between the insult and corresponding punch line. Much can be copied and pasted from before.

Listing 6-2. Using a Grouping Sequence

```
byte randomNum1;      // teacher joke program, second attempt
byte randomNum2;      // Expanded, and even more humorous
boolean counter;
const int trigger = 7;
byte trigState;
byte trigLatch;
String firstPart;
String secondPart;
byte oldRandomNum1;
byte oldRandomNum2;
byte veryOldRandomNum1;
byte veryOldRandomNum2;

void setup(){
  pinMode (trigger, INPUT_PULLUP);
  Serial.begin (9600);
  Serial.println ("tap a wire from pin 7 to ground for a joke");
  Serial.println ();
  Serial.println ();
  randomSeed (analogRead(5));
}

void loop(){
  trigState = digitalRead (trigger);
  if (trigState == LOW){
    trigLatch = 1;
  }
  while (trigLatch == 1){
    do{
      randomNum1 = random (0, 20);
    }
    while ((oldRandomNum1 == randomNum1) || (veryOldRandomNum1 ==
 randomNum1));
// put the line of code shown above on one line
// checks and eliminates redundant phrases in the first

    do{
      if ((randomNum1 < 20) && (randomNum1 > 14)){
                              //picks 19, 18, 17, 16, 15
        randomNum2 = random (15, 20);
      }
```

```
      else if ((randomNum1 < 15) && (randomNum1 > 9)){
                                  //picks 14, 13, 12, 11, 10
        randomNum2 = random (10, 15);
      }
      else if ((randomNum1 < 10) && (randomNum1 > 4)){
                                  //picks 9, 8, 7, 6, 5
        randomNum2 = random (5, 10);
      }
      else {                      //picks 4, 3, 2, 1, 0
        randomNum2 = random (0, 5);
      }
    }
    while ((oldRandomNum2 == randomNum2) || (veryOldRandomNum2
== randomNum2));
          // put the line of code shown above on one line
        // checks and eliminates redundancy in the second part

    veryOldRandomNum1 = oldRandomNum1;     //separates by two
    veryOldRandomNum2 = oldRandomNum2;
    oldRandomNum1 = randomNum1;            //separates by three
    oldRandomNum2 = randomNum2;

    subRoutine1();
    subRoutine2();
    Serial.print ("My teachers are so ");
    Serial.print (firstPart);
    Serial.print (" they ");
    Serial.print (secondPart);
    Serial.println (" ");
    Serial.println (" ");
    trigLatch = 0;
    delay (500);
    trigLatch = 0;
  }
} //end of main loop

void subRoutine1(){
  switch (randomNum1){
  case 19:
    firstPart = "fat,";
    break;
  case 18:
    firstPart = "big,";
    break;
```

```
case 17:
  firstPart = "humongous,";
  break;
case 16:
  firstPart = "large,";
  break;
case 15:
  firstPart = "giant,";
  break;
case 14:
  firstPart = "ugly,";
  break;
case 13:
  firstPart = "homely,";
  break;
case 12:
  firstPart = "nasty lookin,";
  break;
case 11:
  firstPart = "revolting,";
  break;
case 10:
  firstPart = "beastly,";
  break;
case 9:
  firstPart = "stupid,";
  break;
case 8:
  firstPart = "clueless,";
  break;
case 7:
  firstPart = "brainless,";
  break;
case 6:
  firstPart = "dimwit,";
  break;
case 5:
  firstPart = "thick headed,";
  break;
```

```
  case 4:
    firstPart = "cheap,";
    break;
  case 3:
    firstPart = "bargain-basement,";
    break;
  case 2:
    firstPart = "tight with a buck,";
    break;
  case 1:
    firstPart = "low-priced,";
    break;
  case 0:
    firstPart = "stingy,";
    break;
  }
}
void subRoutine2(){
  switch (randomNum2){
  case 19:
    secondPart = "have hooves instead of feet";
    break;
  case 18:
    secondPart = "makes cows look skinny";
    break;
  case 17:
    secondPart = "block the Sunshine";
    break;
  case 16:
    secondPart = "shop for items that contain gluten";
    break;
  case 15:
    secondPart = "put the all-you-can-eat restaurants out-of-
business";
    break;
  case 14:
    secondPart = "buy the extended warranty on mirrors";
    break;
  case 13:
    secondPart = "cause robbers to run the other way and turn
themselves in";
    break;
```

```
case 12:
  secondPart = "cause mirrors to refuse to show a reflection";
  break;
case 11:
  secondPart = "have no bugs that will bite them";
  break;
case 10:
  secondPart = "have the sun hiding behind the clouds when
they're at the beach";
  break;
case 9:
  secondPart = "think little people are inside of the TV";
  break;
case 8:
  secondPart = "think you need to put a computer in the oven,
if it freezes up";
  break;
case 7:
  secondPart = "buy orange juice concentrate to get smarter";
  break;
case 6:
  secondPart = "have the dog teaching them new tricks";
  break;
case 5:
  secondPart = "keep calling the hospital asking for a
hemorrhoid transplant";
  break;
case 4:
  secondPart = "cause bill collectors to send them money";
  break;
case 3:
  secondPart = "try to spend both sides of a dollar bill";
  break;
case 2:
  secondPart = "reuse toothpaste";
  break;
case 1:
  secondPart = "use both sides of toilet paper";
  break;
case 0:
  secondPart = "save leaves in the fall, and paste them back
on in the spring";
  break;
  }
}
```

The changes in this code are not highlighted, but in comparing this latest code for teacher jokes to the code in Listing 6-1, along with an increased number of put-downs, you can see that we are grouping the insults with the punch lines for the jokes to make more sense. Also, we are not allowing the repetition of each section of the joke to occur any closer than a spacing of three. As a project, you can modify the program by both expanding the database of jokes (don't be too mean), and increasing the nonallowance of repeat spacing. Additionally, it is recommended that if you are using this text in a school, you should change the focus of the joke to pass the class with high marks. Lawyer jokes are a good choice, because almost everyone likes lawyer jokes.

Perfecting Random Numbers

In our previous examples of trying to generate discrete responses that were not duplicated, we found that it was possible to limit repetitiveness by spacing the occurrences and pulling from a larger reservoir of possibilities. In this section, we describe simple algorithms that can be used to entirely eliminate the duplication of random numbers. An algorithm is not always a mathematical formula; in coding it is generally a process used to accomplish a goal. It is the logic behind the code. A complex algorithm usually takes a great deal of thought, time, and effort to develop, and cannot be typed into code on the fly. There are a number of methods programmers use to mentally visualize the operation of a program. Some, in fact, do use math, but the majority of programmers will use simple sketches on paper or a more orderly type of drawing called a flowchart, as illustrated in Figure 6-1.

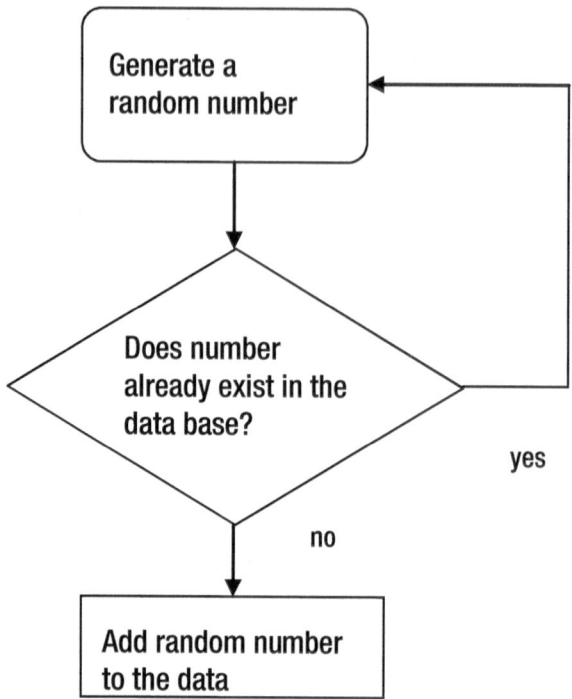

Figure 6-1. *An example flowchart*

Another commonly used method that aids in establishing an algorithm is called pseudocode. This method has you write the code in a somewhat normal language, just as you would speak, but you try to sound a bit like a computer. The code matching the flowchart in Figure 6-1 for not wanting to have duplicated random numbers is shown here as pseudocode:

Generate random number

```
If random number = existing number in data base
Then go back, generate new random number
Else, add new number to data base
```

Flowcharts and pseudocodes are common ways to design program operation, and there are many other ways to devise the logic that is needed to complete the objective. Understanding the objective and taking time to design the logic flow of a program is always time well spent, where in all but the simplest cases, the actual writing of code should be the very last step. A surgeon would have X-rays and MRIs at his or her disposal before operating on a sick patient, an architect would have a blueprint before constructing a building, and an engineer or technician would have a schematic of a device before building it, so it follows that a programmer would have a logic tree and algorithm substantially constructed before beginning to write code.

We have been working with random numbers and have had some issues arise where we wish to eliminate duplicate numbers. The code we generate in Listing 6-3 follows the previous flowcharts and pseudocode for a logical way to eliminate duplication.

Listing 6-3. Eliminating Duplication

```
//This program produces 10 random numbers with no duplication:
int ArrayOne [10];
int ArrayMirror [10];
int i;
int j;
const int trigger = 7;
boolean trigState;

void setup(){
  pinMode (trigger, INPUT_PULLUP);
  randomSeed (analogRead(5));
  Serial.begin (9600);
  Serial.println ("Ground pin 7 for a set of ten different numbers");
}
void loop(){
  trigState = digitalRead (trigger);
  while (trigState == LOW){
    delay (200);
    Serial.println ("Please wait");
dupeRestart:
```

```
//this section sets up your numbers, creates a copy, and prints
    for ( i = 0;   i < 10;   i++ ) {
      ArrayOne [ i ] = random (0, 10);
      ArrayMirror [ j ] = ArrayOne [ i ];
      j ++;
    }
    j = 0;
    for ( i = 0;   i < 10;   i++ ) {  //this section looks for dupes
      for ( j = 0;   j < 10;   j++ ) {
        if ( ( i != j ) && ( ArrayOne [ i ] == ArrayMirror [ j ] ) ) {
          j = 0;
          goto dupeRestart;
// if a dupe encountered, goes to dupeRestart
        }
      }
    }
    //this section prints the array with no dupes
    Serial.println (" ");
    for ( j = 0;   j < 10;   j++ ){
      Serial.print (ArrayMirror [ j ] );
      Serial.println (" ");
    }
    trigState = HIGH;
  }
}
```

This code works fine for very small sets of numbers, up to about 10 or so, when using the Arduino UNO. The processing time becomes too great, however, when larger sets of numbers must be generated because if any number in the set is not unique, the code throws out the entire set and restarts the process from the beginning. The result will be correct, but the iterations of an entire set of numbers could take quite a long time as the set gets long. We fix this issue in our next project, so let's now look at an interesting way of checking for duplicate numbers. The code uses two for loops, with the algorithm based on the mechanical logic of two separate circles that hold identical numbers. Each circle holds what is termed an *array.* Arrays of numbers use one variable but have many possibilities for discrete values contained in separate instances of the array that are differentiated by an index number assigned to each instance of the one variable. At the top of our code we declare two arrays: `ArrayOne` holds 10 discrete values indexed from 0 to 9, and `ArrayMirror` is the second array that equally is declared to hold 10 discrete values also indexed from 0 to 9. In our logic, we consider the outside circle mentioned earlier to be `ArrayOne`, and the inside circle to be `ArrayMirror`.

Please visualize this as you examine the code shown in the main loop: The outer circle clicks clockwise only one click as the inner circle rotates 360 degrees. The equal index numbers are ignored, but if any other instance of an equal value exists between the two arrays, it indicates a duplicate. In this way, each outer number of the slowly rotating circle values is compared to each and every inner circle value. If a duplicate is noted, the random number generation for the arrays will reoccur. The code in Listing 6-4 addresses the elapsed time problem we had in our previous code that threw out the entire set of numbers if there was a duplicate. Now, by identifying the nonduplicate numbers and storing them in a final array, only the duplicate numbers need be regenerated, as needed.

Listing 6-4. Eliminating the Elapsed Time Problem

```
int ArrayOne [10]; //produces 10 nonduplicate numbers efficiently
int ArrayMirror [10]; //by adding to the nonduplicate array
int ArrayDupes [10]; //progressive iterations
int i;
int j;
int d;
const int trigger = 7;
boolean trigState;
boolean goToReset;

void setup(){
  pinMode (trigger, INPUT_PULLUP);
  randomSeed (analogRead(5));
  Serial.begin (9600);
```

```
  Serial.println ("Ground pin 7 to start");
  Serial.println (" ");
}


void loop(){
  trigState = digitalRead (trigger);
  delay (100);
  while (trigState == LOW) { //this section sets up your numbers,
                          //creates a copy, and prints number
    Serial.println (" ");
    Serial.println ("Original Numbers ");
    Serial.println (" ");
    for (i = 0; i < 10; i++) {
      ArrayOne [i] = random (1, 11);
      ArrayMirror [j] = ArrayOne [i];
      Serial.print ( ArrayOne [i] );
      Serial.println (" ");
      j++;
    }
reset:                  //restarts here to clear duplicate numbers
    goToReset = false;

    for ( d = 0; d < 10; d++ ) {  //zeros the dupe array
      ArrayDupes [d] = 0;
    }

    for ( i = 0; i < 10; i++ ) {
      for ( j = 0; j < 10; j++ ) {  //tests for dupes

        if ( (ArrayOne [ i ] == ArrayMirror [ j] ) && ( i >  j)
&& ( i !=0 )) {

          ArrayDupes [i] = ArrayOne [i]; // gives array of dupes
                                     // gives zeros for none
        }
      }
    }
    i=0;
    for ( d = 0; d < 10; d++ ) { //if ArrayDupes numbers are dupes
                          // picks a new random number
```

```
  if ( ArrayDupes [ d ] != 0 ) {
    ArrayOne [ d ] = random (1, 11);
    goToReset = true;
  }
}
j =0;
for ( i = 0;  i < 10;  i++ ) {
  ArrayMirror [ j ] = ArrayOne [ i ];
  j++;
}
if (goToReset == true) {
  goto reset;              // since there were dupes, goes back to
                           //run dupe test again
}
Serial.println (" ");
Serial.println ("Non-Duplicate Numbers ");
Serial.println (" ");
for (i = 0;  i < 10;  i++ ) { //prints an array of no dupes
  Serial.print ("   ");
  Serial.print ( ArrayOne [ i ] );
  Serial.println (" ");
}
Serial.println (" ");
    i = 0;
    j = 0;
    d = 0;
    for ( i = 0; i < 10; i++ ) { //zeros the ArrayOne array at end
                                 //of the game
      ArrayOne [ i ] = 0;
    }
    trigState = HIGH;
  }
}
```

There might be other more efficient methods to generate arrays of nonduplicate numbers, but the code as shown builds a new nonduplicate array of a distinct amount of numbers that can very easily be used to simulate a deck of cards containing 52 distinct outcomes. We utilize this code in later sections of the text, as we build card games to have fun with programming.

Poker Game

We will now combine the nonduplicative random number code with more use of the array process to play the game of five-card straight poker. In this game, five cards are dealt to the player and five to the dealer. The rules of poker apply to determine which hand is better. We did not incorporate the rules of the game into this code, as our main objective is now simply to demonstrate an application where the nonduplicative random number code is useful. A more complete version of this game is shown later in the book. As an extended project, you can add code to the extended game shown later to display the player or dealer as the winner of the hand. We must keep in mind however, that the onboard memory on the UNO is limited to 2 kB, and that a microcontroller's real job is to examine inputs and produce corresponding outputs. The advantage of working with the Arduino microcontroller comes from its open source hardware and software: It has tremendous popularity in the maker space, as well as a high degree of standardization. There are many project applications and add-on modules available, called *shields*. Shields plug in over the top of the Arduino board and increase its functionality. There are even breadboard type shields available that allow the user to construct hardware device input and output on an attached circuit board that sits atop the Arduino. A controller's job is to do useful things, and we are misusing it by playing games.

In the following program, some of the code is the same as that for our last project. We have highlighted the new lines and changes. For consistency between the grouping of cards and the different suits, such as diamonds, hearts, and so forth, we have formatted them into Table 6-1.

Table 6-1. *Sorting Cards*

Suit 0 = Clubs	Suit 1 = Diamonds	Suit 2 = Spades	Suit 3 = Hearts
Card 1 = Ace	Card 21 = Ace	Card 41 = Ace	Card 61 = Ace
2	22	42	62
through	through	through	through
10	30	50	70
Card 11 = Jack	Card 31 = Jack	Card 51 = Jack	Card 71 = Jack
Card 12 = Queen	Card 32 = Queen	Card 52 = Queen	Card 72 = Queen
Card 13 = King	Card 33 = King	Card 53 = King	Card 73 = King

Because we have card numbers 1 through 13 accounted for in the first column, our algorithm was to disregard the section of card numbers until the next number's least significant digit matched. That is why it was determined that card 21 would be the ace in the next suit. It is slightly inefficient to generate random numbers that are not in use, but the processing time lag is not seriously affected, and the coding is more straightforward. Listing 6-5 is the program that follows from Table 6-1.

Listing 6-5. Five-Card Poker Game

```
int ArrayOne [72];      //5 card poker with no dupes
int ArrayMirror [72];
int ArrayDupes [72];
int ArrayDeal [52];
int ArrayPlayer [5];
int ArrayDealer [5];
int suit;
int card;
int x;
int i;
int j;
int d;
const int trigger = 7;
boolean trigState;
boolean goToReset;
```

```
void setup(){
  pinMode (trigger, INPUT_PULLUP);
  randomSeed (analogRead(5));
  Serial.begin (9600);
  Serial.println ("Ground pin 7 to deal");
  Serial.println (" ");
}

void loop(){
  trigState = digitalRead (trigger);
  delay (100);
  while (trigState == LOW){
    Serial.println (" ");
    Serial.println ("Please Wait");
    for (i = 0; i < 72; i++){
      ArrayOne [i] = random (1, 74);
      ArrayMirror [j] = ArrayOne [i];
      j++;
    }
reset:                    //restarts here to clear duplicate numbers
    goToReset = false;

    for (d = 0; d < 72; d++){   //zeros the dupe array
      ArrayDupes [d] = 0;
    }

    for (i = 0; i < 72; i++){
      for (j = 0; j < 72; j++){
        if ( (ArrayOne[i] == ArrayMirror [j]) && (i > j) && (i != 0
)){ //dupes test
          ArrayDupes [i] = ArrayOne [i]; //array of dupes and zeros
                                         //for none
        }
      }
    }
    i = 0;
    for (d = 0; d < 72; d++){       //if numbers are dupes,
                                    //picks a new number
      if (ArrayDupes [d] != 0){
        ArrayOne [d] = random (1, 74);
        goToReset = true;
      }
    }
    j =0;
    for (i = 0; i < 72; i++){ //loads mirror array
```

```
    ArrayMirror [j] = ArrayOne [i];
    j++;
  }
  if (goToReset == true){
    goto reset; // since dupes, goes back to run the dupe test again
  }
  for (i = 0; i < 52; i++){ //Dupes are gone, grabbing the first
                           //10 good random numbers generated
      if ((ArrayOne [i] < 14) || ((ArrayOne [i] >20) && (ArrayOne [i]
< 34)) || ((ArrayOne [i] > 40) && (ArrayOne [i] < 53)) || (ArrayOne
[i] > 60)) {
        ArrayDeal [x] = ArrayOne [i];
        x++;
      }
  }
  Serial.println (" ");
//now test print of original random numbers-remove to play normally
  for (x = 0; x < 10; x++){
    Serial.println (ArrayDeal [x]);
  }

  Serial.println (" ");
  for (x = 0; x < 10; x++){

    if (ArrayDeal [x] > 60){
      suit = 3;
      card = ArrayDeal [x] - 60; //this normalizes the card value
    }
    else if (ArrayDeal [x] > 40){
      suit = 2;
      card = ArrayDeal [x] - 40;
    }
    else if (ArrayDeal [x] > 20){
      suit = 1;
      card = ArrayDeal [x] - 20;
    }
    else if (ArrayDeal [x] < 14){
      suit = 0;
      card = ArrayDeal [x];
      //card = card
    }
```

```
Serial.println (" ");
if (x == 0) {
  Serial.println ("Player's Cards:");
  Serial.println (" ");
}
if (x == 5){

  Serial.println (" ");
  Serial.println ("***************");
  Serial.println (" ");
  Serial.println ("Dealer's Cards:");
  delay (2000);
  Serial.println (" ");
}

if (card == 1){
  Serial.print ("The Ace");
}
else if (card == 11){
  Serial.print ("The Jack");
}
else if (card == 12){
  Serial.print ("The Queen");
}
else if (card == 13){
  Serial.print ("The King");
}
else{
  Serial.print (card);
}

Serial.print (" of ");

switch (suit) {
case 3:
  Serial.print ("Hearts");
  break;
case 2:
  Serial.print ("Spades");
  break;
case 1:
  Serial.print ("Diamonds");
  break;
case 0:
  Serial.print ("Clubs");
  break;
  }
```

```
  }
  Serial.println (" ");
  x = 0;
  i = 0;
  j = 0;
  d = 0;
  for (i = 0; i < 72; i++) { //zeros ArrayOne array at end of game
    ArrayOne [i] = 0;
  }
  trigState = HIGH;
  }
}
```

Multidimensional Arrays

In the past few projects, we used arrays to keep track of individual values contained in number sets. Occasionally, more complex analysis of groupings might be needed; that is when two- and three-dimensional arrays can really come in handy. The project code in Listing 6-6 illustrates a two-dimensional array. We define two small arrays with three possibilities each, using the subscripts 0, 1, and 2. Having two combinations of three possibilities equals a total of nine possible outcomes ($3^2 = 9$). Each value is loaded, and on running the program, the serial monitor will display each of the distinct values associated with the two arrays.

Listing 6-6. A Two-Dimensional Array

```
int myArray [3][3]; //each dimension has 3 possibilities 0, 1, and 2
int i;
void setup() {
  Serial.begin (9600);
}
void loop() {
  myArray[0][0] = 1; //loads array locations
  myArray[0][1] = 2;
  myArray[0][2] = 3;
  myArray[1][0] = 4;
  myArray[1][1] = 5;
  myArray[1][2] = 6;
  myArray[2][0] = 7;
  myArray[2][1] = 8;
  myArray[2][2] = 9;

  while (i < 1) { //runs only one time
    Serial.println (myArray[0][0]); //prints values in the array
    Serial.println (myArray[0][1]);
    Serial.println (myArray[0][2]);
    Serial.println (myArray[1][0]);
    Serial.println (myArray[1][1]);
    Serial.println (myArray[1][2]);
    Serial.println (myArray[2][0]);
    Serial.println (myArray[2][1]);
    Serial.println (myArray[2][2]);
    i++;
  }
}
```

Dice Game

The following dice game is a great learning example of how to write computer code that must produce specific results for a given set of circumstances. In engineering and technology, we must find solutions to problems. The problem given to us in this section is to write code for a computer to simulate throwing a pair of dice. The game has simple rules that readily lend themselves to computer coding, but at the same time provide a caveat in that if no win or loss is achieved in the first round, a completely different set of rules then supersede the first set of rules, until the game starts over.

The first set of rules that are in effect for the first throw are as follows: Two dice are thrown (each is called a die), and if the dot markings on both dice add up to the numbers 7 or 11, then the person throwing the dice is a winner. If on the first throw any one of the following numbers result, however, the throw is considered to be a loss: 2, 3, or 12. The caveat is that if the number of dots resulting from the first throw is neither a win nor a loss (i.e., 4, 5, 6, 8, 9, 10), then the number that is thrown is called the point. If a point is the result of the first throw, then the player must repeatedly throw the dice in hopes of matching the point number to win the game. If in the process, however, the player's dot pattern thrown equals the number 7, then he or she loses the game, and the game resets back to the initial set of rules for the next throw. In coding this process, the first set of rules and the second set of rules are both straightforward. The complication is moving between the two sets of rules. One method to solve this problem is shown in Listing 6-7.

Listing 6-7. Using Two Sets of Rules

```
int randomNum1;
int randomNum2;
const int trigger = 7;
int trigState;
int trigLatch;
int roll;
int point;
int counter;

void setup(){
  pinMode (trigger, INPUT_PULLUP);
  Serial.begin(9600);
  Serial.println ("Ground pin 7 to roll");
  Serial.println (" ");

  randomSeed(analogRead(5));
}
```

```
void loop(){
  trigState = digitalRead(trigger);
  if (trigState == LOW){
    trigLatch = 1;
  }
  while (trigLatch == 1){
    randomNum1 = random(1,7); //gets a random 1 through 6 for die one
    Serial.println (randomNum1);
    delay (700);

    randomNum2 = random(1,7); //gets a random 1 through 6 for die two
    Serial.println (randomNum2);
    delay (700);

    roll = (randomNum1 + randomNum2); //get total
    delay (300);
    Serial.print ("You rolled ");
    Serial.println (roll);
    delay (700);
    trigLatch = 0;
    // FIRST TIME THROUGH COUNTER IS ZERO//
    if (counter == 0){
      if (roll == 7 || roll == 11){     //7 or 11 win
        Serial.print (roll);
        Serial.println (" is a Winner");
        Serial.println ("");
        delay (800);
        counter = 0;
        roll = 0;
        trigLatch = 0;
      }
      if (roll == 2 || roll == 3 || roll == 12){ //2, 3, or 12 lose
        Serial.println ("You Lose");
        Serial.println ("");
        delay (800);
        counter = 0;
        roll = 0;
        trigLatch = 0;
      }
```

```
      if (roll == 4 || roll == 5 || roll == 6 || roll == 8 ||
roll == 9 || roll == 10){
        point = roll;
        Serial.println ("Roll again to hit the point");
        delay(400);
        counter++;
        roll = 0;//needed to negate second code section on first roll
        trigLatch = 0;
      }
    }
    //SECOND TIME THROUGH COUNTER > ZERO//
    if (counter > 0){
      counter++;
      if (roll == point){
        Serial.println ("You Win, You hit the Point");
        Serial.println ("");
        delay (700);
        counter = 0;
        roll = 0;
        trigLatch = 0;
      }
      if (roll == 7){
        Serial.println ("You lose");
        Serial.println ("");
        delay (700);
        counter = 0;
        roll = 0;
        trigLatch = 0;
      }
      if (counter == 4){ //helpful comments to the user
        delay (200);
        Serial.print ("**You need ");
        Serial.println (point);
        delay (700);
        trigLatch = 0;
      }
```

```
if (counter == 7){
  delay (200);
  Serial.print ("**Looking for ");
  Serial.println (point);
  delay (700);
  trigLatch = 0;
}
if (counter == 10){
  delay (200);
  Serial.print ("**This is taking awhile, You need ");
  Serial.println (point);
  delay (700);
  trigLatch = 0;
}
  }
  }
}
```

The preceding code used a series of multiple if statements. The if statement is a very powerful conditional statement that allows for the decision-making process of a computer program. Here is an analogy: Traveling on the Ohio Turnpike on the way to Cleveland, there was a road sign spotted saying "Cleveland Left." Some people might turn around and return home because they would think that Cleveland was not there anymore, whereas an autonomous vehicle with the computer code if Cleveland then exit would arrive at the destination.

Review Questions

1. How can you enable spell check in the Arduino IDE?

   a.   Press Ctrl+S.

   b.   Type SPELL CHECK in all capital letters.

   c.   Choose Spell Check from the menu.

   d.   There is no spell check.

2. A quick way to format code is to

   a.   type FORMAT in all capital letters.

   b.   choose Format from the menu.

   c.   Press Ctrl+T.

   d.   Choose Format in the Options box.

3. A graphical way to plan for program operation is to use

   a.   pseudocode.

   b.   functions.

   c.   loops.

   d.   flowcharts.

4. A while loop will run continuously while a condition exists. (True/False)

5. A do-while loop will execute a block of code how many times at the absolute minimum?

6. Which of these is a more efficient way of performing many if conditional statements?

   a.   Switch case

   b.   do loops

   c.   if then, as if

   d.   Breaks

7. Which of these is a way of planning a program using words that are similar to actual code?

   a.   Flowchart

   b.   Pseudocode

    c.   Conditional statements

    d.   Conditional testing

8. `For` loops are identical to `while` loops but run only once. (True/False)

9. What is an algorithm?

10. What is the geometric shape for a decision in a flowchart?

    a.   Diamond

    b.   Square

    c.   Parabola

    d.   Triangle

Project 6

Add to the database by including more insults to the humorous teacher joke program. (Don't be too outlandish, or they might flunk you.)

CHAPTER 7

More Game Programming, with a Detailed Explanation

Coding the Game 21: First Attempt

We will continue creating interactive games from scratch. They are simplistic video games, as modern computer and gaming consoles have such lifelike graphics, animation, and sounds that even the military is using the technology to train solders for combat scenarios. We, however, are essentially using a device only meant to be a controller of equipment to provide us with a little entertainment as we learn the fundamentals of coding. Our next program is the card game 21, also called blackjack. Card games work well for learning how to code because the games have specific rules and outcomes. We have a player and dealer, each given two cards. The cards' numerical values range from 2 up to 10, and there are four picture cards, for a total of 13 possibilities. We will code a random number of 2 to 14, which gives us all of the possibilities. The code for this is `"random (2, 15)"`. This first version of the game is incomplete, but it begins the design process. Other more complete versions of the game are provided later. This first iteration does not use nonduplicate numbers and does not address the point value of picture cards; we will do that later. For now, we review concepts in a slow, methodical process and begin to develop the game. Open the Arduino IDE and load the code shown in Listing 7-1 without statements followed by the double forward slashes (//), as they are comment lines only seen by other programmers and not recognized by the processor as code. Comments are important for you and others to understand the thought process of the programmer. When writing your own code, please use them. Also, do not type in the line numbers as they are shown for illustration purposes only.

Listing 7-1. Coding for 21 Game

```
1 const byte triggerPin = 7;
2 int randomNumbers [4]; //declares an array of 4 values,(0 thru 3)
3 boolean trigState;
4 boolean trigLatch;
5 byte counter;

6 void setup(){        // this section sets I/O, runs only once
7 pinMode (triggerPin, INPUT_PULLUP);
8 Serial.begin (9600);
9 Serial.println ("Ground pin 7 to start");
10 Serial.println ();
11 randomSeed (analogRead(5));
}

12 void loop(){              // Main Program Loop Begins:
13 trigState = digitalRead (triggerPin);
14 if (trigState == LOW){
15 trigLatch = true;
   }
16 while(trigLatch == true){ //trigLatch = false stops this loop
17 getNumbers(); //calls the function/subroutine getnumbers
18 for (counter = 0; counter < 4; counter++){
19 Serial.println (randomNumbers[counter]); //prints numbers
20 delay(1000);        //the delay makes it interesting to watch
   }
```

```
21 trigLatch = false;
Serial.println ();
}               //end of trig latch
}                   // end of main loop

 // Functions, subroutines, and Interrupts are below,
//outside of the main program:

 void getNumbers(){
    for (counter = 0; counter < 4; counter++){
    randomNumbers[counter] = random(2, 15);
    }
 }
```

The code for our program is numbered to aid in providing a complete description. Old programming languages relied on line numbering, but used a trick of not making the line numbers consecutive, and would instead increment by 10 or more, just in case a programmer needed to make future changes and insert code between the lines. It is good to note that the processor does indeed follow each line of code from the top of the code to the bottom (unless redirected). The old Basic programming language, as well as others, allowed for subroutines to keep things orderly and decrease repetitious sections of code. The idea was to have the main program go to a separate section outside of the step-by-step procedure to perform a task, and then return at the point in the main program where it left off. Most of today's modern programming languages do not recommend the use of subroutines, but a similar concept called a function can be employed with the Arduino to accomplish the same goal. In our sample code, we use a function to place the section of code used to generate a set of random numbers and then store them in an array.

Referring to the top section of code in our program, lines 1 through 5 are where variables are declared and can be initialized. Remember that in algebra, a variable is a letter that can represent a number. In coding, we can use more than one letter to represent a single number, and instead of using something very abstract like *xyz*, we want to name a variable using a group of letters that are somewhat descriptive. This helps the programmer and others reading the code to better understand what the variables represent. In our example, the variables are `triggerPin`, `randomNumbers`, `trigState`, `TrigLatch`, and `counter`. Notice a variable must consist of a letter or group of letters without a space. It is not necessary, but it is common practice to capitalize the first letter of the second word if you are using a variable that needs more than one word to be descriptive. Some programmers like to instead use an underscore to make the words readable. Using the underscore method, our variable `triggerPin` could be called `trigger_pin`. Either method works just fine to allow for good human readability of the code. There is no spell check when you are writing code, which could lead to a debugging headache in large programs. The processor does not care what you call a variable, as long as you are consistent throughout your program. The processor does, however, need to know what type of number and how much space in computer memory to allocate to each variable. Because we are only referring to a header pin number on the Arduino board with the `triggerPin` variable, it is byte size. In computers, the *binary system* is used, where a bit is a single memory cell that can hold a 1 or a 0, and eight memory cells are called a byte, so a byte is 8 bits. Its maximum size would be if the 8 bits were 1s, and that equates to the number 255 in our normal decimal system. We have about 20 header pins that can be used as I/O on the board, so it is fine to use the byte designation for their memory size, and because the pin number will not be changing, we call it a constant. In algebra, it is permissible to let a letter represent a constant. It is standard practice in math to use the beginning letters of the alphabet for constants and the end of the alphabet for variables that will change values as they are manipulated. For program coding purposes, we only care about naming them for ease of human readability. The first line of our code is telling the processor from that point forward, anytime we refer to the term `triggerPin`, we are specifying header pin 7. This occurs after we identify it as I/O in the setup section. In electronics, the term *trigger* specifies a signal of short duration that begins or ends an operation; for example, a push button switch turning a television on or off. In our code, I chose pin 7 as the start pin because 7 is my lucky number. You could have just as well called it START or anything else that is descriptive, as long as you are consistent throughout the program. You could have also assigned any other appropriate I/O header pin to perform the trigger function. Please note that in this language, uppercase and lowercase letters do matter, so the variable START is not the same as start; they would be entirely separate variables. Also, the code const byte has to be spelled exactly in that way, and in that order, or you would generate a syntax error and the program would not compile. When a program compiles, the computer language is changed into 1s and 0s that can be loaded into the computer memory to run the program. Many times the IDE will highlight the error and try to explain a naming problem, but the Arduino IDE is not perfect and cannot find the exact problem 100% of the time. Common problems to watch out for are neglecting to end each sentence with a semicolon (;). The following are also commonplace errors in syntax: misspellings, errors in capitalization, word and letter spacing, and—the hardest to notice—a curly brace misplacement. The curly braces start and stop a block of code that works as a cohesive unit. The braces look like this: { starts a section, } stops the section. Sometimes the program logic requires you to have sections within sections. To keep the braces from getting confusing, many good programmers will immediately put both the open and the close brace on the page, and then fill out the code in between.

Earlier we discussed generating random numbers through the use of an array; there are a great many ways to write a program where the methodology could be completely different but accomplish the same objective. It would be just as possible to generate each random number in our program rather than doing it in one fell swoop. The use of arrays is a very efficient but slightly difficult concept to understand. It is a fast way to group values as a specific block of data. There is one variable with many different values. Each specific data value is given a variable subscript called an index number.

In line 2 of our code, we are specifying integers using the syntax `int`. Integers are also sometimes called short and variables assigned with this designation use 2 bytes of computer memory space and can contain decimal numbers between –32,768 and 32,767. Integers also are whole numbers without a fractional part. If you needed greater accuracy in computations, you could use the data type called *float,* which allows for fractional values. In our code, we are generating numbers between 2 and 14 to represent the distinct values in a deck of cards, with card numbers 11 through number 14 translated to equal the jack through ace cards. In line 2 the number inside the brackets gives the maximum number of discrete values for the array. We could have also more appropriately used byte as the type. Notice the brackets ([ ]) are different from the braces we discussed earlier. Brackets are used for arrays, whereas braces ({ }) are used for enclosing sections of code operations. Because we specified the number 4 within our brackets, there will be four distinct values, indexed from 0 to 3. We need four cards because the player and dealer get two cards each. Notice that computers start counting at zero, so zero through three accounts for four distinct values. For example, if you specifically wanted to reference the second card generated, you would use the code `randomNumbers[1]` because index zero is card 1, and index 1 is card 2. Yes, it is a little confusing, but computers start counting at zero. After a little gnashing of the teeth, we schedule a trip to the dentist and our frown turns to a clean and happy smile, as we get use to starting with zero.

When we declare a variable outside of the program, it is a called a *global* or *general declaration* and will be available for use anywhere. For efficiency, it is possible to also use a variable in a specific area of the code and in this case, it is called *local*. After declaring the global variables in the top section, in lines 1 through 5 of the code, we proceed to the setup area where we identify pins as being input or output. The Arduino has the ability of using the data lines in a bidirectional manner. This means that a sensor, such as a switch, could be connected to a specific pin as an input to the controller, or an actuator such as an LED could be connected as an output using the same pin; however, a specific pin can only be used as either an input or output at any given time. We can easily touch a small wire from our input pin to the USB metal box on an Arduino board to get a ground (which is a computer *low*). The `pullup` designation on line 7 means that without any connection, the pin is a computer level *high*, as it is pulled up. We want to display real-time data on our computer monitor so we enable serial communication in line 8. The `(9600)` in the parentheses specifies the communication speed in baud, which is bits of data exchange per second; this is a very slow speed. Before broadband Internet connections, analog telephone modems were commonly 56 K baud, or nearly six times as fast, but in our Arduino programs, we are only interested in exchanging text between a computer and a controller and 9600 baud is fast enough. Program line 9 prompts the user, and 10 provides a line space. The `randomSeed` command line appearing next in the setup section elects to pick up electrical interference commonly called noise from the analog pin 5, which is designated an input. Electrical noise is everywhere and caused by power lines, radio and television stations, and even naturally occurring events in outer space. Usually in electronics we try to shield against noise, but in our generation of random numbers it will help the randomization process because the noise is random and unpredictable.

Now to where the magic happens: Line 12 is the main section of the program. It is called a loop because it loops around to the bottom of the section continually while power is applied. The loop might be broken by a function or subroutine for a time, but returns when the external process is completed. The top variable designation section and the setup section of the code only run once when power is first applied or the program is first loaded or reset, but the main loop keeps running over and over. The term `void` just means that the loop is a function that returns no specific value. Going back to algebra again to look at functions, the formula $y = x^2$, is a function that produces a parabola. If you made a graph and picked numbers of both positive and negative values as the domain (x), the range (y) would consist of a parabolic object centered about the positive y axis with its origin at zero. In computer languages, functions usually produce a result such as a number to coincide with the meaning of functions of mathematics; however, for a controller such as the Arduino, we do not usually want to have a result to a math problem. We instead want to sense things and do things. Because of this difference, just about every function in controller programming is a void, which means that we are not getting an overall number result. Just as with a computer starting a count at zero, eventually our frowns turn to happy smiles as we get used to it.

Looking at program lines 13 through 16, the main program looks for a trigger low on pin 7, which we called `triggerPin` (this is when we momentarily tap a wire from pin 7 to the USB metal ground). It momentarily stores the information if a low is there in a memory location we called `trigState`. Then it remembers by storing the fact into a variable called `trigLatch`. `trigLatch` is being used as a longer term memory. When it is true (1), the program will run until we release it by setting `trigLatch` to its original state of false (0). In general terms in electronics, a latch locks in a condition until it is reset. A push button on a television remote is momentary. When you push the on button on your remote the television latches onto that command, until you push a button again and trigger it back to the off state. Again, keep in mind when writing your own code, you have the opportunity to name variables anything you want, but it is best to be descriptive so that other people can figure out your thought process. In addition, you might need to revise your own code sometime in the future and might not remember the logic that you were using. This also leads to the important subject of documentation. You will notice statements in the code that we are using that start with a double backslash (//). They are comment lines, as we mentioned before, and the computer ignores them, but they help us explain our thought process from one human to another.

To generate our random numbers in the main loop section, this code calls a function on program line 17 that we named `getNumbers`. The function is essentially a subroutine residing outside of the main loop. Again, it is a `void` function because it really does not produce an actual result, but it does generate an array of four random numbers. Each number will be between 2 and 14, representing the card values in a deck. It accomplishes the generation through the use of a `for` loop in the function or subroutine. The `for` command institutes a loop that ends after a condition is met. This generates a distinct random number for each spin around the function loop. The `counter` variable is incremented each time through by the syntax command `counter++`, which adds one every time it executes. After the four times around the loop, the program pointer goes back to the spot in the main program loop where it left off, and continues to execute the remainder of the main loop code. The next section in the main loop, program lines 18 through 20, uses another `for` command to print the random numbers that were generated to the computer monitor. After all the numbers are displayed on the monitor, the `trigLatch` is broken at program line 21. The main program continues to loop but ignores the `trigLatch` interior code until trigger pin 7 is made low once more.

Coding the Game 21: Second Attempt

In the next area of code, we want to associate picture cards with their values. We are assigning 11 to the ace because that is the value of points, and we go on to assign 12 to the jack, 13 to the queen, and 14 to the king. After allowing the players to know what card they were dealt, though, we have to make their card game point value equal to 10 for the face cards other than the ace. (In the sample code that follows in Listing 7-2, this procedure is done only for the player. To make the game functional, you will need to complete the procedure for both of the dealer's cards. Use copy and paste as much as possible and just change the specific values.)

Listing 7-2. Another Version of the 21 Game

```
//second version of 21, picture cards ok for player not dealer yet
const byte triggerPin = 7;
int randomNumbers [4];//this declares an array of 4 values(0 thru 3)
boolean trigState;
boolean trigLatch;
byte counter;
byte yourTotal;
byte dealerTotal;

void setup(){
  pinMode (triggerPin, INPUT_PULLUP);
  Serial.begin (9600);
  randomSeed (analogRead(5));
  Serial.println (" ");
  Serial.println ("Ground pin 7 to play");
  Serial.println (" ");
}
void loop(){                              //main program loop
  trigState = digitalRead(triggerPin);
  if (trigState == LOW){
    trigLatch = true;
  }
  while(trigLatch == true){//to exit this loop make trigLatch = false
getNumbers();
Serial.println (" ");
Serial.println ("****NEW GAME****");
Serial.println (" ");
delay (1000);
Serial.print (" Your First Card is ");
switch (randomNumbers [0]){
  case (11):
    Serial.print ("an Ace worth ");
    break;
    case (12):
    Serial.print ("a Jack worth ");
    break;
    case (13):
    Serial.print ("a Queen worth ");
    break;
```

```
      case (14):
      Serial.print ("a King worth ");
      break;
}
if ((randomNumbers [0]) > 11){
   (randomNumbers [0]) = 10;
}
Serial.print (randomNumbers [0]);
delay (1500);
Serial.print (", Your Second Card is ");
switch (randomNumbers [1]){
   case (11):
     Serial.print ("an Ace worth ");
     break;
     case (12):
     Serial.print ("a Jack worth ");
     break;
     case (13):
     Serial.print ("a Queen worth ");
     break;
     case (14):
     Serial.print ("a King worth ");
     break;
}
if ((randomNumbers [1]) > 11){
   (randomNumbers [1]) = 10;
}
Serial.println (randomNumbers [1]);
delay (2000);
Serial.println (" ");
Serial.println (" ");
Serial.print (" The Dealer's First Card is ");
Serial.print (randomNumbers [2]);
delay (1500);
Serial.print (", The Dealer's Second Card is ");
Serial.print (randomNumbers [3]);
delay (2500);
yourTotal = randomNumbers [0] +  randomNumbers [1];
dealerTotal = randomNumbers [2] +  randomNumbers [3];
Serial.println (" ");
Serial.println (" ");
Serial.println ("                                        ");
Serial.print (" Your total is ");
```

```
        Serial.println (yourTotal);
        Serial.println (" ");
        delay (2000);
        Serial.print (" The Dealer's Total is ");
        Serial.println (dealerTotal);
        Serial.println (" ");

    trigLatch = false;

}            //end of trig latch
}                        // end of main loop

///// this is the function used to generate the random numbers /////

void getNumbers(){
    for (counter = 0; counter < 4; counter++){
    randomNumbers[counter] = random(2, 15);
    }
}
```

Review Questions

1. Subroutines are frowned on in C, C++, and the processing language that the Arduino code is based on, but _____ can be implemented to do pretty much the same thing.

   a. arrays

   b. functions

   c. floats

   d. brackets

2. Arrays in Arduino code use _____ to contain
   the index.

   a. { } curly brackets

   b. [ ] solid brackets

   c. *\ asterisk and backslash

3. Why should variables be somewhat descriptive?

4. In older computer languages, lines of code were
   numbered. (True/False)

5. The integer data type contains _____, whereas
   the float contains _____.

6. Why are arrays useful?

7. The control structure very frequently used with
   arrays is

   a. floats.

   b. if then statements.

   c. do loops.

   d. variables.

8. In Arduino code "xyz" would be

   a. one variable.

   b. two variables.

   c. three variables.

   d. a constant.

9. Using the statement `Serial.print ("c")`, which of the following would occur?

   a. A carry operation would occur.

   b. A carry operation will print.

   c. A letter c would print to the serial monitor.

   d. A space would print to the serial monitor.

10. In making variables descriptive, how many spaces between variable words are allowed?

   a. zero

   b. one

   c. two

   d. three

Project 7

Rewrite the code for the game of 21 using different variables.

CHAPTER 8

Electronic Projects

The circuits presented in this chapter require basic electronic components readily available at online retail outlets. Links can be found for complete kits and parts vendors from the author's official website at www.dukish.com.

Coding a Voltmeter

As discussed in Chapter 1, LEDs have a voltage drop of approximately 2 volts across their internal junction, and the voltage can vary depending on the size, color of the LED, and forward current. We will use an analog input on the Arduino and measure the exact value of the voltage drop. Because the microcontroller is digital, the analog information must be converted into the binary system to be processed. In general, when we go from analog to digital we use a device called an analog-to-digital converter (ADC). Going the other way, coming from the digital realm to the analog world, we use a digital-to-analog converter (DAC). For us to measure an analog voltage, as opposed to a logic level, the Arduino ADC will convert the voltage to a binary number between 0 and 1,023. The number 1,023 is the maximum voltage that the processor can handle, which equates to 5 volts with the UNO. If you are consistently measuring a voltage lower than 5 volts, use the analog reference pin AREF to increase the ADC accuracy. A reference voltage can easily be developed across a resistor voltage divider. Our project code, shown in Listing 8-1, uses the entire range and

© Bob Dukish 2018
B. Dukish, *Coding the Arduino*, https://doi.org/10.1007/978-1-4842-3510-2_8

reads the voltage once each second, and then displays the ADC value between 0 and 1,023, equating to the actual voltage value of between 0 and 5 volts. For finding the actual voltage from the ADC, in our case we divide 5 by the 1,024 steps for the conversion factor (5/1,024 = 0.004883), then the actual voltage is found by multiplying the ADC step number reading, by the conversion factor, which is the height of each step. Because we want a precise voltage value, rather than using the integer data type, we declare variables as float for floating point decimals. Voltage is also described as a difference of potential, and measured across a component. Because our Arduino voltmeter is referenced to ground potential, though, we can only directly check across the bottom component in a circuit, so we will build the LED circuit shown in Figure 8-1 and connect our analog input pin between the LED and the resistor. We happened to pick pin A0 as our input pin, but the UNO has six analog pins, and any of them could have been used. They can also be used for digital I/O when needed.

Listing 8-1. Coding a Voltmeter

```
const int voltageIn = A0; //Makes a volt meter up to 5 volts
int voltLevel;
float actualVoltage;
void setup (){
  Serial.begin(9600);
  pinMode (voltageIn, INPUT);
}
void loop(){
  voltLevel = analogRead(voltageIn);
  actualVoltage = (voltLevel * .004883); //volt steps are .004883
  Serial.println();
  Serial.print("The level is ");
  Serial.println (voltLevel);
  Serial.print("The actual voltage is ");
  Serial.println (actualVoltage);
  Serial.println();
  Serial.println();
  delay(1000);
}
```

After running the code for the voltmeter, the ADC number representing 2 volts is 410 with both numbers being displayed on the serial monitor if you were measuring a typical red LED; however, the values could vary due to the type of LED and component tolerances. Using Kirchhoff's voltage law, we can assume that because 5 volts is the total voltage, and we were reading 2 volts across the LED, there must be 3 volts across the current-limiting resistor. We can also measure the voltage across the resistor, but because the Arduino is taking a reading with ground as the reference point, we must interchange the LED and the resistor in the circuit. After swapping the two components around and running the program, you should observe approximately 3 volts across the resistor. As one last measurement, we could reverse the LED and measure zero volts across the resistor. Because the LED and resistor are in a series circuit and the current loop is broken, no current will flow through the resistor to produce a voltage drop across it (remember Ohm's Law). The current is near zero because the LED is a polarized component and the current can only flow in one direction through a diode. There is a negligible amount of reverse current, but if the input voltage were to drastically increase, a breakdown of the LED junction could occur and cause a short circuit. There is no voltage across a short circuit and a maximum voltage across an open circuit. We can simulate an open LED by orienting the two components into their initial positions as shown in Figure 8-1. Then with the resistor on top and the LED on the bottom, reverse the polarity of the LED by flipping around so that the negative side is facing toward the positive voltage source. The LED should be off and the voltmeter will read 5 volts across the open circuit. This illustrates two valuable indications for troubleshooting circuits: Shorts have zero volts across them, and there will be maximum voltage across an open circuit. With the LED reverse biased, it is essentially an open point in the circuit.

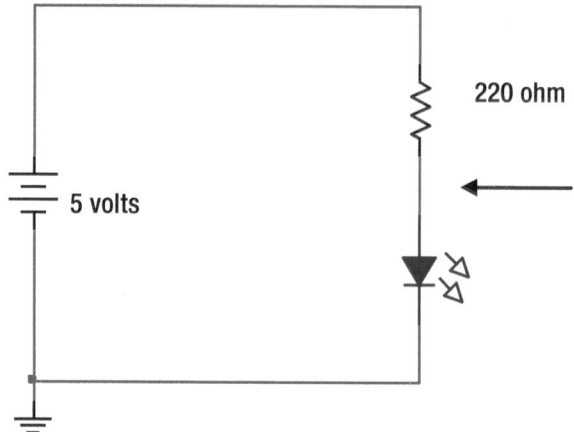

Figure 8-1. *An LED circuit*

Dimming an LED with Pulse Width Modulation

In the last project, we used an analog voltage as an input to the microcontroller. This would be useful in a variety of applications where analog sensors are used to either produce a varying voltage or resistance as conditions change. In this project, we will produce a pseudoanalog output coming from the microcontroller. Although a digital device can only approximately generate a true analog output, because there are voltage level steps involved in the DAC process just as we saw with the ADC, the steps can be made small so as to approach a true analog signal output. There are two ways to produce an analog output: One way is to actually produce a varying voltage with the DAC, and the other is by varying the pulse widths of square waves. In a few of our earlier projects we rapidly flashed an LED and noted that its brightness appeared steady but dim. This pseudoanalog technique is called pulse width modulation (PWM).

In digital electronics, there are two logic levels. For the Arduino UNO, a low level is zero and a high is 5 volts. A high pulse is shown between the two arrows in Figure 8-2, with a low pulse shown immediately to the right. Both the high pulse and low pulse make up one cycle. In Figure 8-2, we would say that the signal is a square wave because the pulse widths are equal. The time on is equal to the time off. It has a 50% duty cycle. When the times of the cycles are short, and the flashing is fast, our eyes do not discern the flashes, but instead we will notice that the overall brightness decreases. The characteristic of our vision called *persistence* allows us to watch television and movies and not notice any flickering between frames. If the high pulse time shown in Figure 8-2 were to lessen, the duty cycle would go down and the LED would appear dimmer because the entire cycle time would remain the same but the LED would be on for a smaller fraction of the total. This explains how PWM works with vision, but PWM can also be used as a way to modulate communication signals and to control motor speed.

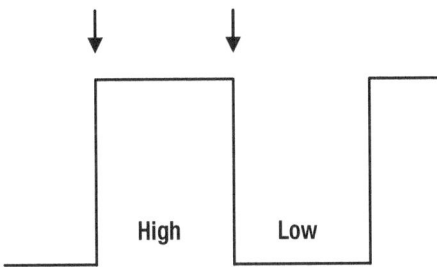

Figure 8-2. *Pulse width modulation*

The Arduino has a number of pins for PWM; on the UNO they are 3, 5, 6, 9, 10, and 11. This project uses PWM to control the brightness of an LED. Start by connecting a 220 Ohm resistor and an LED between pin 9 and ground. After loading the code in Listing 8-2, we can control the LED intensity by momentarily grounding any one of three pins. We use pin 7 for 100% duty cycle (full brightness), pin 6 drops to medium brightness, and pin 5 drops the LED to low brightness. The command `analogWrite` identifies the output and the number that follows is the PWM duty cycle broken up into 256 steps, 0 to 255, with 255 being the highest duty cycle producing full output. We picked the medium value to be 50% duty cycle, and the low to be about 25% duty cycle.

Listing 8-2. Dimming an LED with PWM

```
const int LED = 9;    // A three intensity LED program
const int highPin = 7;
const int medPin = 6;
const int lowPin = 5;
boolean high;
boolean med;
boolean low;
void setup(){
  pinMode (LED, OUTPUT);
  pinMode (highPin, INPUT_PULLUP);
  pinMode (medPin, INPUT_PULLUP);
  pinMode (lowPin, INPUT_PULLUP);
}
void loop(){
  high = digitalRead (highPin);
  med = digitalRead (medPin);
  low = digitalRead (lowPin);
  if (high == LOW){
  analogWrite(LED, 255);
  }
 if (med == LOW){
  analogWrite(LED, 125);
  }
  if (low == LOW){
  analogWrite(LED, 60);
  }
  delay(100);
}
```

Controlling an LED Using a Light Sensor

In the schematic in Figure 8-3, we use a photo resistor to vary the conduction of an NPN transistor circuit used to illuminate an LED.

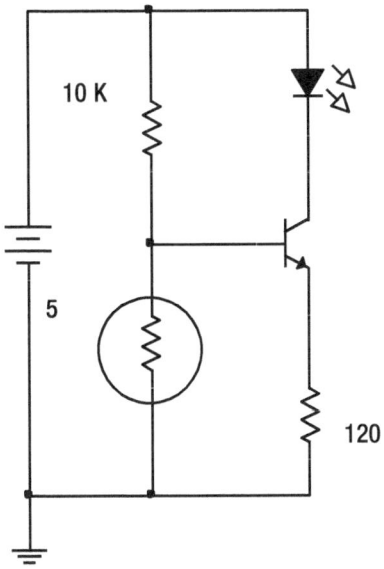

Figure 8-3. *Transistor circuit used to illuminate an LED*

The photo resistor, as shown in the circle, exhibits decreasing resistance as light increases. There is a voltage at the point between the two resistors that also connects to the control element of the NPN transistor, called the base. If there is decreasing resistance across the photo resistor caused by increasing light, the corresponding decreasing voltage also appears on the transistor base. If the base voltage decreases, it produces less base current through the transistor, which causes the LED to have less current through the vertical section of the transistor's emitter and collector sections, which therefore causes the LED to go dimmer, or off (i.e., more ambient light, less LED light). Conversely, if there is less light on the photo resistor, the resistance goes higher, the voltage goes higher, and the current through the LED goes higher, causing it to get brighter (i.e., less ambient light, more LED light). The objective is that in bright ambient light the LED is off, and in low light the LED is on. If the light changes are somewhat gradual on the photo resistor, the electronic circuit will produce a somewhat gradual change in LED intensity. If parts are available, build and test the circuit.

We will use the Arduino to switch an LED on and off. This exercise (Listing 8-3) could easily be adapted to control outdoor lighting, security systems, and other devices for which operation is dependent on differentiating day from night. To add the microcontroller, connect a wire from the intersecting point between the two resistors and transistor base to the Arduino analog input pin A0. We are using an analog input because the voltage developed across the photo resistor will vary in an analog manor related to the amount of light intensity. (You might also wish to later modify the program to output a PWM signal to vary the LED brightness, but our program is only interested in sensing between light and dark, and correspondingly switching an LED off or on. Please modify the code to achieve a different switching response, and you might need to do this, as ambient light conditions will vary.) The analog voltage developed across the photo resistor is represented by a number of from 0 to 1,023. We could scale our input for more accuracy by using an external reference connected to the AREF pin on the Arduino, and adjust the code accordingly. In our example, we just use the ADC number of 1,024 to represent the full 5 volts. (Again, keep in mind that you might need to make adjustments to the analog read numbers we call lowLight in Listing 8-3, depending on the amount of light intensity you are working in.)

Listing 8-3. Controlling an LED with a Light Sensor

```
const int LED = 13;
int lowLight;
boolean on;
void setup(){//analog pins are inputs by default
  pinMode (LED, OUTPUT);
  Serial.begin(9600);
  Serial.println("the serial monitor is displaying the light value");
}
void loop (){
  lowLight = analogRead (0); //reads voltage at 1024 steps
  if (lowLight > 140) { //ambient light low, turn on LED
    digitalWrite (LED, HIGH);
    on = true;
  }
  if ((lowLight <= 140) && (on == false)){
    digitalWrite(LED, LOW); // ambient light high, turn off LED
  }
  if ((lowLight <= 120) && (on == true)){     //hysteresis
    digitalWrite(LED, LOW);
  }
  Serial.println (lowLight);
  delay (500);
}
```

In the bottom section of code where we commented about hysteresis, without adding that section, the controller could have possibly flickered the LED when the ambient light was very near the switching threshold. Hysteresis is used to lock in a function until there is a large change in the input. It is used in thermostats so that the heating or cooling periods are distinct, so that temperature control units do not repeatedly cycle on and off near the set point.

Analog sensors tend to be a little finicky, so the code that we presented might need to be adapted to your specific lighting conditions and breadboard circuit build. That is why we added the serial monitor function into the code. After uploading the code, you can open the serial monitor and check the analog read as you expose the photo resistor to light and dark conditions, and then adjust your numbers for the proper switching function. Once the Arduino code is working properly, you will notice that the transistor circuit varies the intensity of its LED in an analog manor, whereas the controller abruptly switches the onboard LED near pin 13 on and off.

Analog circuits can also abruptly switch logic levels. This can be done in several ways, with one solution being the use of a hybrid device called a *comparator,* which is basically an open-ended operational amplifier (op-amp), which is configured as a high-gain voltage amplifier. With small changes near the switching point it can jump from rail to rail, between low and high levels. Comparators are available as ICs. On the other hand, microcontrollers can simulate analog outputs, as with the PWM project that we looked at in the last section. With the inclusion of a DAC they can mimic analog circuit output, as in the case of a CD or MP3 player. There are gray areas between analog and digital technology.

Coding a Frequency Counter

The schematic shown in Figure 8-4 was first presented in Chapter 4 to demonstrate how a process could be implemented either by using discrete physical components or by creating a program to have a microprocessor perform the process. The process that we are examining is a free-running astable multivibrator. Due to the resistive and capacitance components connected to the NE555 (LM555) timer, a square wave output is observed on IC output pin 3, which flashes an LED once a second. (The output is approximately one half-second off and one half-second on.) Timing circuits like this are useful in providing clock pulses for timing purposes; however, the accuracy of a 555 timer circuit is not very good due to the wide tolerances in resistive and capacitance components. Common resistors have a plus or minus tolerance of 5%, and capacitors have an even wider tolerance. If accuracy were an issue, a hardware solution could be a crystal controlled oscillator, and such ICs are readily available.

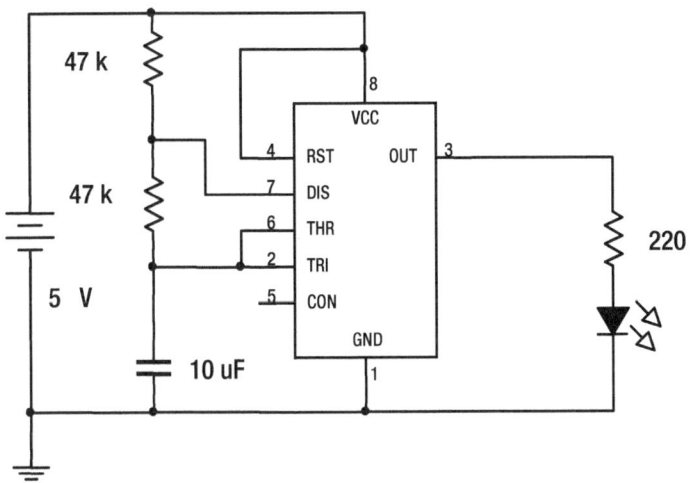

Figure 8-4. *A timing circuit*

Our next project is to use the 555 timer IC as a square wave frequency source and create a program to read its frequency and create an output on the Arduino indicating that the frequency is within tolerance. If the electronic components are not available, you can use a function generator or a second Arduino to act as the frequency generating device, as described in the next section.

To power the device, we are connecting the VCC power line of the 555 to the 5-volt header pin on the Arduino to use USB power, and the ground line to one of the Arduino ground pins. Pin 3 of the 555 should be connected to pin 3 on the Arduino; the LED circuit can remain connected on the breadboard. We use pin 3 on the Arduino just because of the nice number match, but any digital I/O pin will work just as well (see Figure 8-5).

Figure 8-5. *Powering the device*

The breadboard has a gap in the middle separating two similar sides. The holes on each side running parallel with the short edge of the breadboard are all connected. Usually there are five holes in a row on each side, and they are connected together, but are not connected to the holes on the other side of the gap, or to any other holes. Running perpendicular along the long side of the breadboard are two parallel lines of holes. Each parallel line running along the length of the board is connected, but they are not connected anywhere else. These two sets of parallel lines are manly for use as power buses and can each be jumped together to the sets of lines on the other side of the board, which can be seen on the far right side of Figure 8-5. The IC is placed in any convenient location straddling the middle gap. It can be seen in Figure 8-5 that the breadboard is getting 5-volt power and ground from the Arduino, which is connected to a computer via the USB connector. As mentioned, also connect a wire between pin 3 of the IC on the breadboard and pin 3 on the Arduino.

If built as shown in Figure 8-4, the 555 will generate one pulse per second, which equates to a frequency of 1 Hz. The code in Listing 8-4 will read the square wave pulses from the Arduino and display the number 1 on the serial monitor, showing that it is counting a 1 Hz signal.

Listing 8-4. A Frequency Counter

```
//freq counter reads sine, square, and triangle waves
unsigned long currentMillis;    // works from 1 Hz to 20 Khz
unsigned long lastMillis;       //samples once a second
unsigned long duration;
int pulseHigh;
int pulseLow;
long halfCycles;
long cycles;
int in = LOW;
void setup(){
  pinMode(3, INPUT);
  Serial.begin(9600);
  Serial.println("The frequency is displayed in Hertz");
}

void loop(){
  lastMillis = millis();
  do{
    in = digitalRead(3);
    currentMillis = millis();
    duration = currentMillis - lastMillis;

    if (in == HIGH){
      pulseHigh = 1;
    }
    if (in == LOW){
      pulseLow = 1;
    }
    if (pulseHigh ==1 && pulseLow == 1){
      halfCycles = halfCycles + 1;
      pulseHigh = 0;
      pulseLow = 0;
    }
  } while(duration < 1000); //looks for cycles per second

  cycles = halfCycles / 2; //divide by two, there are two half cycles
                           //in a cycle
  Serial.println(cycles);
  cycles = 0;
  halfCycles = 0;
  delay (1000);
}
```

169

The do-while loop in the code is using the millis timer to count an elapsed time of 1 second, as frequency is defined as cycles per second (Hz).

The pulse count is triggered as the waveform goes through one alternation between positive and negative value, as shown in Figure 8-6, and then again for the low pulse between the negative going transition and the next positive going transition. Because this trigger point accounts for one half of the waveform and occurs twice for each cycle, the number is divided by two to report the actual frequency in cycles per second (Hz). The code will work for square waves, sine waves, and triangle waves. If a function generator is available, it would be an interesting project to try all three waveforms at different frequencies up to 20 kHz. Care must be taken, however, so that the waveforms are in the range between 0 and 5 volts DC. Use of a TTL output is convenient.

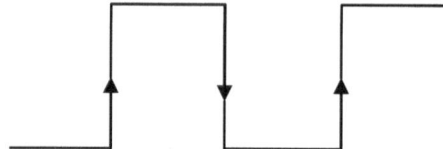

Figure 8-6. *Alternation between positive and negative values*

We now change our 555 circuit's frequency by removing and replacing the 10 µF capacitor with a 0.1 µF capacitor to increase the frequency from about 1 Hz to 100 Hz. The 0.1 µF cap is nonpolarized, meaning that there is no polarity consideration for positive and ground. Small capacitors under 1 µF are generally nonpolarized. Because of their small physical size, a code is sometimes used to identify the value of small capacitors. The first number represents the first digit, followed by the second digit, with the third number representing the number of zeros, and the total value as picoFarad in engineering notation. Our 0.1 µF cap will have the code 104, as 1 and a 0 followed by four more 0s means 100,000 pF, and that is equal to 0.1 µF. (If you have a question about this, be sure to review the engineering notation information earlier in the text.)

The adaptation to the previous program will display the exact frequency of the 555 on the serial monitor as before; however, the

code is now slightly enhanced to additionally flash the onboard LED
connected to pin 13, if the frequency is within plus or minus 10 Hz of
100 Hz. Lighting the LED could also be used in a real-world application of
checking for a good signal condition. Additionally, the code, or the circuit,
could be modified to flash a lamp or sound a buzzer if the signal goes
out of tolerance. The changes to the code to now respond to the correct
frequency are shown as highlighted in Listing 8-5.

Listing 8-5. Responding to the Correct Frequency

```
unsigned long duration;
unsigned long currentMillis;
unsigned long lastMillis;
int pulseHigh;
int pulseLow;
long halfCycles;
long cycles;
int in = LOW;
const int LED = 13;
void setup(){
  pinMode(3, INPUT);
  Serial.begin(9600);
  pinMode(LED, OUTPUT);
}
void loop(){
  lastMillis = millis();
  do{
    in = digitalRead(3);
    currentMillis = millis();
    duration = currentMillis - lastMillis;

    if (in == HIGH){
      pulseHigh = 1;
    }
    if (in == LOW){
      pulseLow = 1;
    }
    if (pulseHigh ==1 && pulseLow == 1){
      halfCycles = halfCycles + 1;
      pulseHigh = 0;
      pulseLow = 0;
    }
```

171

```
} while(duration < 1000);      //looks for cycles per second

cycles = halfCycles / 2;

Serial.println(cycles);
if ((cycles > 90) && (cycles < 110)){
digitalWrite (LED, HIGH);
}
cycles = 0;
halfCycles = 0;
delay (1000);
digitalWrite (LED, LOW);
}
```

Pulse Generation

The Arduino can produce a frequency output, as was done previously with the 555 timer IC. The advantage of using code to control the frequency is that no IC, resistors, or capacitors need to be used. Our code also allows for real-time control of the frequency via text input to the serial monitor. The frequency output is also displayed on the serial monitor and sent to a digital output pin to control an LED and a normal speaker, or possibly a piezo speaker. A piezo speaker uses the piezoelectric effect to generate sound. The piezo element is made of a crystalline material that makes sound as a varying voltage is placed across the element. It does this by distorting (bending) as the voltage varies. The piezoelectric effect has reciprocity, as do most devices that act as a *transducer*. A transducer is a device that converts one form of energy to another. In our case, the element will convert electrical energy to mechanical energy that will produce sound.

As a side note, we all know the law of conservation on mass and energy, which basically states that matter and energy cannot be created or destroyed, but can be changed from one form to another. Once we use electrical energy to bend a crystal and produce sound, or move a speaker diaphragm, then, what form of mass-energy does the process ultimately produce? (We will save that thought for a chapter review question.)

As mentioned, the piezoelectric transducer has reciprocity, which means that it bends when a voltage is applied across its element, and conversely it also can produce electricity when a mechanical force is used to bend it. It could be used as either an input device (sensor) or an output device (actuator), as in our application. One advantage of using a piezo speaker is that they are inexpensive and lighter in weight than a normal speaker. Piezoelectric elements also have a capacitive effect and capacitive devices pass higher frequencies better, so the higher frequencies will generally have better sound fidelity than will low frequencies. Typical human hearing has a range of from 20 Hz to 20,000 Hz (20 kHz), although as one ages the frequency response to high frequencies is lessened (some people say that this accounts for long marriages). The lowest frequency we can produce with our code is 35 Hz, due to the `tone` command limitation. On the high side, we can produce frequencies well above the range of human hearing. Because we are using the unsigned `int` type for the `tone` command (Listing 8-6), we can go as high as 65,535 Hz. The piezo speaker should give the most favorable results between 500 and 5,000 Hz. If you do not have a piezo speaker, you can connect a small normal 8 Ohm speaker to the Arduino. As shown in Figure 8-7, the current must be limited by adding the series resistor as shown, or damage to the Arduino could result. We are also using the onboard LED near pin 13, so you can see the results even without a sound-producing device. Even at the lowest Arduino tone of 35 Hz, the individual flashes are not distinguishable, but the LED flashing results in reduced brightness. You will also notice that when an "x" is input in the serial monitor, the LED completely extinguishes because the pulse generation output stops.

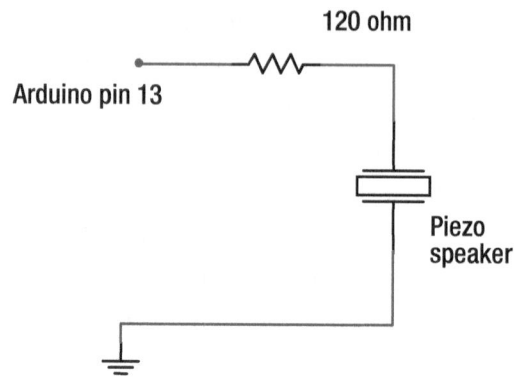

120 ohm

Arduino pin 13

Piezo
speaker

Figure 8-7. *Adding a series resistor*

Listing 8-6. Entering Frequency for Pulse Generation

```
//code to enter frequency for pulse generation

unsigned int inputNumbers;
unsigned int frequency;
String numbers = "";            // empty string for input numbers
const int LED = 13;

void setup() {
  pinMode (LED, OUTPUT);
  Serial.begin(9600);
  Serial.println("This program is for pulse generation");
  Serial.println();
  Serial.println("Enter a number from 500 hz up to 5000 hz");
  Serial.println();
  Serial.println("(You must put hz after the number, and then
press enter or click send.)");
  Serial.println();
}
void loop() {
  while (Serial.available() > 0) {    // user keys in data
    inputNumbers = Serial.read();
    if (isDigit(inputNumbers)) {
      numbers += (char)inputNumbers;
      //puts the numbers in sequence as string, converts to numbers
    }// now looks for the upper or lowercase ASCII code for h:
```

```
    if ((inputNumbers == 104) || (inputNumbers == 72)) {
      Serial.print("You entered a pulse frequency of: ");
      Serial.print(numbers.toInt());
      Serial.println(" hz");
      Serial.println("re-enter if incorrect, to end enter x");
      Serial.println();
      frequency = (numbers.toInt());
      tone (LED, frequency);
      //clears variables
      inputNumbers = 0;
      numbers = "";
    }
  if ((inputNumbers == 120) ||   (inputNumbers == 88)){
      // upper or lower case x, shuts off tone function
      inputNumbers = 0; //clears variables
      numbers = "";
      Serial.println("idle, enter frequency for output:");
      Serial.println();
      noTone (LED);
    }
  }
}
```

We look for numbers from the serial monitor input and ignore other characters while determining the frequency that is to be used with the `tone` command. The numbers are put together as string type data and then converted to an integer, when the user inputs either letter h or H for Hertz. When the user inputs x or X, the output stops. This project outputs a series of pulses at a given frequency. Square waves have tremendous distortion and sound terrible as audio signals, but the pulses we are producing in our project could have many different applications beyond that of just producing sound.

Counter with Seven-Segment Display (with Driver IC)

It is possible to use the Arduino to directly light the seven-segment display, but the next two projects reduce wiring by using a 74LS47 BCD to seven-segment display driver TTL IC. Figure 8-8 shows the pin-out for the seven-segment display. It is oriented with its left side on the bottom, and the display's decimal points shown to the right. It has the same numbering scheme as a 14-pin IC. Shown with the segments facing the viewer, pins 1 through 7 run left to right along the bottom, and then proceed counterclockwise to pins 8 through 14 along the top, which run right to left. (Please note that pins 4, 5, and 12 are missing on the display, and this also aids in finding the proper orientation.) The segment letters shown connect to the pins as illustrated in the schematics for the projects that follow. Additionally, 5 volts connects through a current-limiting resistor to common anode pins 3 and 14. The projects presented save time and minimize parts by only using one current-limiting resistor connected on the anode side of the display. This, however, will cause fluctuations in the overall brightness as different numbers are displayed. Ideally, a current-limiting resistor should be placed in each of the cathode sections. In using a common anode device, a low on the pin associated with a segment will cause the segment to light. (A high on the pin will not allow for conduction through the LED segment, and it will not light.)

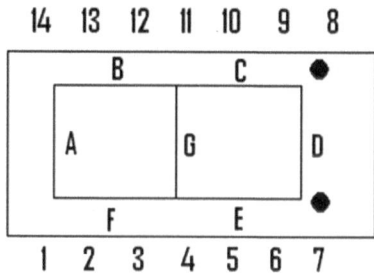

Figure 8-8. *MAN 72 common anode seven-segment display*

The schematic (Figure 8-9) and code (Listing 8-7) for this project will cause the seven-segment display to show each number between 0 and 9, after Arduino pin 8 is momentarily touched to ground. After the number 9 has been displayed, the count returns, and holds at 0 until Arduino pin 8 is retriggered. The count can be reset during the operation, however, if Arduino pin 7 is momentarily grounded. A similar, but more complex circuit is used in the next project. It would be best to build this project and preserve it, keeping in mind that a similar additional display and two switching transistors will be added in Listing 8-8. A picture of the breadboarded circuit is shown in that section (Figure 8-10).

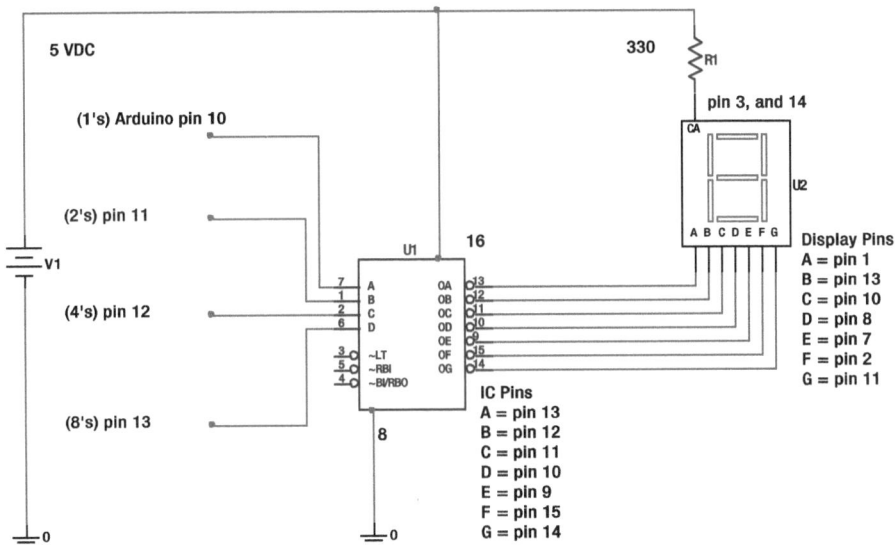

Figure 8-9. *The schematic for this project*

Listing 8-7. Seven-Segment Display

```
  /*This program causes a seven-segment display with a driver
IC to count to 9
 * The code will reset the count to 0, or pin 7 will reset
during the count
 * As an exercise, modify the code to only count up to the number 5
 */
int maxCount;
const int trigger = 8;
const int setReset = 7;
int trigState;
int trigLatch;
int timer;
boolean action;
void setup() {
  pinMode (trigger, INPUT_PULLUP);
  pinMode (setReset, INPUT_PULLUP);
  pinMode (10, OUTPUT);
  pinMode (11, OUTPUT);
  pinMode (12, OUTPUT);
  pinMode (13, OUTPUT);
}
void loop() {
  trigState = digitalRead (trigger); //looks for ground to start
  if (trigState == 0) {
    trigLatch = 1;
    maxCount = 1; //starts the count at 1
  }
  while (trigLatch == 1) { //starts counting
    while (maxCount < 10) { //resets and stops after number 9
      do {
        timer = timer + 1;
        action = digitalRead (setReset); //looks for reset pin press
        if (action == LOW) {  //if reset pin 8 is pressed
          maxCount = 0; // resets to 0 and continues to count to 9
        } //ends if
        switch (maxCount) {     //selects the number to display
          case 1:
            digitalWrite (10, HIGH);  //displays 1
            break;
```

```
    case 2:
      digitalWrite (11, HIGH);
      break;
    case 3:
      digitalWrite (10, HIGH);
      digitalWrite (11, HIGH);
      break;
    case 4:
      digitalWrite (12, HIGH);
      break;
       case 5:
      digitalWrite (10, HIGH);
      digitalWrite (12, HIGH);
      break;
    case 6:
      digitalWrite (11, HIGH);
      digitalWrite (12, HIGH);
      break;
    case 7:
      digitalWrite (10, HIGH);
      digitalWrite (11, HIGH);
      digitalWrite (12, HIGH);
      break;
    case 8:
      digitalWrite (13, HIGH);
      break;
    case 9:
      digitalWrite (10, HIGH);   //displays 9
      digitalWrite (13, HIGH);
      break;
   }
   delay (10); //displays zero and resets display
   digitalWrite (10, LOW);
   digitalWrite (11, LOW);
   digitalWrite (12, LOW);
   digitalWrite (13, LOW);
  } while (timer < 100);  //creates a do loop time to be visible

  maxCount = maxCount + 1;
  timer = 0;         //resets loop timer
}                    //exits this while loop when maxCount = 10
```

```
    trigLatch = 0; //resets the trigger latch, ready for new trigger
    maxCount = 0;
    timer = 0;
  }
}
```

Dice Game with Seven-Segment Display (with Driver IC)

This microprocessor-based game, shown in Figure 8-10, randomly generates two numbers that simulate a dice roll. The NPN switching transistor is a 2N3904. It provides a current path to +5 volts (VCC) to the display, as the 74LS47 IC driver provides a ground for the respective segments to light. The 330 Ohm resistor limits the current flow. If the display is too dim, replace the resistor with a 220 Ohm. (Note that we have slightly changed the output pins on the Arduino; now the least significant digit is pin 11, and the most significant digit is pin 13. Pin 10 on the Arduino is not used, and pin 6 of the driver IC is wired to ground.) The schematic in Figure 8-11 shows one of the displays; however, you must construct a second identical seven-segment display connected in parallel with the 74LS47 IC seven-segment driver outputs, and also construct a second 5-volt switching transistor circuit. Two displays are needed to display both dice numbers. Because only one display driver IC is necessary, the best way to connect the second display is to jump a wire from the driver output side of each LED segment connection in the first circuit over to the second display. The seven segment wires act essentially as a data bus, and the transistor circuit provides a chip enable for the appropriate device. This type of bus design is very common in computer hardware.

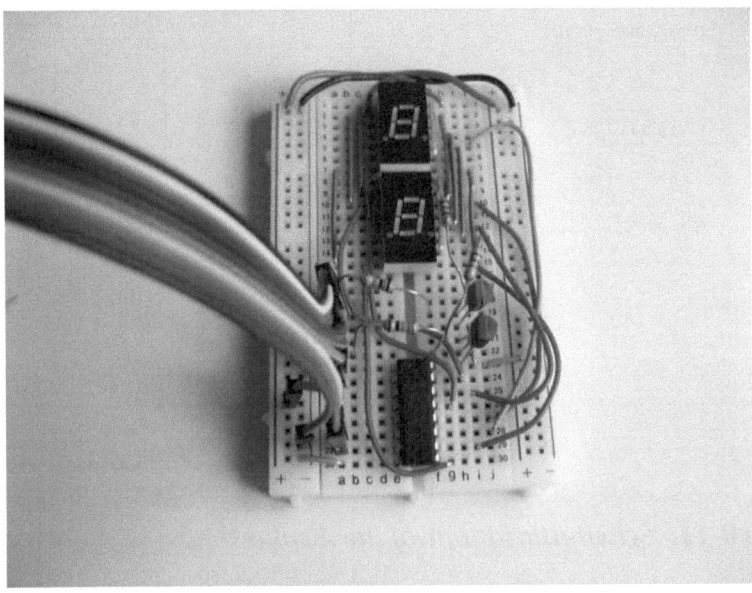

Figure 8-10. *Seven-segment display for dice game*

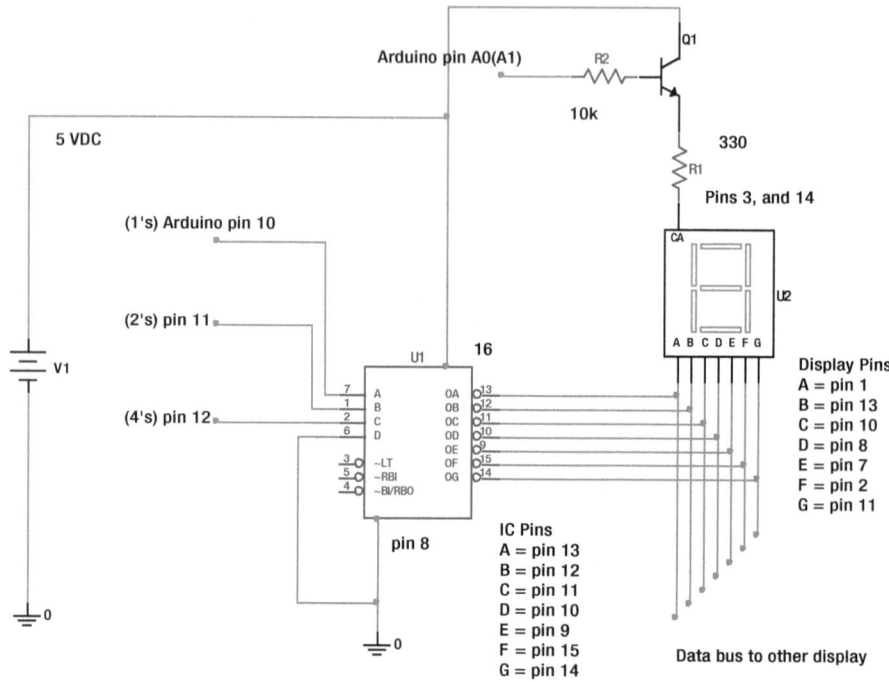

Figure 8-11. *Schematic for one of the displays*

The code in Listing 8-8 selects one of the two displays by providing a high to the base of the transistor, thus switching it on to provide current from 5-volt VCC to display a number. After a very short time period, we select the other transistor switching circuit to enable the other display to show a different number. In this way, we are able to *multiplex* the data to minimize the number of Arduino output ports, as well as minimize power draw. Essentially the displays will alternately flash both numbers as they are rapidly selected and as the appropriate data are output on the BCD bus. The flashing is much too fast for the human eye to discern, and it will appear as though two separate numbers are being displayed simultaneously. The use of this multiplexing technique will also be necessary for later projects. To simplify the code in this section, we eliminate code used for the serial monitor. This project allows you to see the result of a dice roll without the IDE serial monitor. Later, we will connect LEDs in place of the displays to simulate the dots on a pair of dice, and code for wins and losses.

Listing 8-8. Code for the Dice Game Display

```
int randomNum1;
int randomNum2;
const int trigger = 8; //pin 8 initiates the dice roll
int trigState;
int time;

void setup(){
  pinMode (trigger, INPUT_PULLUP);
  pinMode (11, OUTPUT);
  pinMode (12, OUTPUT);
  pinMode (13, OUTPUT);
  pinMode (A0, OUTPUT);
  pinMode (A1, OUTPUT);
  randomSeed(analogRead(5));
}

//looks for button press, generates and displays two random numbers
void loop() {
  trigState = digitalRead (trigger);
  if (trigState == 0){
  randomNum1 = random(1, 7);
  randomNum2 = random(1, 7);

  do { //sequentially displays each of the two numbers
  time = time + 1;
  digitalWrite(A0, HIGH); //selects die one switching transistor
  digitalWrite(A1, LOW);

  switch (randomNum1){ //selects LEDs to light for the random numbers
                    //generated 1 through 6
    case 1:
    digitalWrite (11, HIGH); //binary 1
    break;
```

```
  case 2:
  digitalWrite (12, HIGH); //binary 2
  break;
   case 3:
  digitalWrite (11, HIGH); //binary 1
  digitalWrite (12, HIGH); //binary 2, total is 3
  break;
   case 4:
  digitalWrite (13, HIGH); //binary 4
 break;
    case 5:
 digitalWrite (13, HIGH);   //binary 4
 digitalWrite (11, HIGH);   //binary 1, total is 5
 break;
    case 6:
 digitalWrite (13, HIGH);   //binary 4
 digitalWrite (12, HIGH);   //binary 2, total is 6
 break;
}
delay (10); //keeps first display on 10 ms in loop
digitalWrite (11, LOW);
digitalWrite (12, LOW);
digitalWrite (13, LOW);
digitalWrite (A0, LOW);
digitalWrite (A1, HIGH); //selects die two switching transistor

  switch (randomNum2) {
 case 1:
  digitalWrite (11, HIGH);
  break;
  case 2:
  digitalWrite (12, HIGH);
  break;
   case 3:
  digitalWrite (11, HIGH);
  digitalWrite (12, HIGH);
  break;
   case 4:
  digitalWrite (13, HIGH);
  break;
```

```
    case 5:
  digitalWrite (13, HIGH);
  digitalWrite (11, HIGH);
  break;
    case 6:
  digitalWrite (13, HIGH);
  digitalWrite (12, HIGH);
  break;
}
 delay (10); //keeps second display on 10 ms in loop
  digitalWrite (11, LOW);
  digitalWrite (12, LOW);
  digitalWrite (13, LOW);
  digitalWrite (A0, LOW);
  digitalWrite (A1, LOW); //shuts off both displays
  } while (time < 200); //back to do, until end of 4 second loop
trigState = 1;           // resets trigger to high
time = 0;                //resets time
} //ends if
} //ends loop
```

Counter with Seven-Segment Display (No Driver IC)

This section uses the Arduino to drive each LED segment of the display directly. Driving the display directly not only eliminates the IC, but also allows for special effects. The drawback is that more Arduino ports are needed and the wiring complexity is increased. There is no schematic, but the wiring is explained in the code section (Listing 8-9). It is helpful to refer to the pin-out of the MAN72 and some of the circuitry shown in the schematic of Figure 8-9 and Figure 8-11.

Listing 8-9. Using Arduino to Drive Each LED Segment Directly

```
const int A = 1; //pin names to connect to each segment of the display
const int B = 13;
const int C = 10;
const int D = 8;
const int E = 7;
const int F = 2;
const int G = 11;
const int trigPin = 6;
int trigLatch; //variables
boolean trigState = HIGH;
int maxCount;
int timer;

void setup() {
  pinMode (A, OUTPUT);   //display segments
  pinMode (B, OUTPUT);
  pinMode (C, OUTPUT);
  pinMode (D, OUTPUT);
  pinMode (E, OUTPUT);
  pinMode (F, OUTPUT);
  pinMode (G, OUTPUT);
  pinMode (trigPin, INPUT_PULLUP);
}
void loop() {
  digitalWrite (A, LOW); //lows light up, this starts with zero
  digitalWrite (B, LOW);
  digitalWrite (C, LOW);
  digitalWrite (D, LOW);
  digitalWrite (E, LOW);
  digitalWrite (F, LOW);
  digitalWrite (G, HIGH); //high disables the middle segment
  trigState = digitalRead (trigPin); //looks for low on pin 8 to start
  if (trigState == 0) {
    trigLatch = 1;
  }
  while (trigLatch == 1) { //starts counting
    while (maxCount < 10) { //resets and stops after number nine
      do {
        timer = timer + 1;
        switch (maxCount) {
```

```
case 1:
  digitalWrite (A, HIGH); //segments low light up
  digitalWrite (B, LOW);
  digitalWrite (C, LOW);
  digitalWrite (D, HIGH);
  digitalWrite (E, HIGH);
  digitalWrite (F, HIGH);
  digitalWrite (G, HIGH);
  break;
case 2:
  digitalWrite (A, LOW);
  digitalWrite (B, LOW);
  digitalWrite (C, HIGH);
  digitalWrite (D, LOW);
  digitalWrite (E, LOW);
  digitalWrite (F, HIGH);
  digitalWrite (G, LOW);
  break;
case 3:
  digitalWrite (A, LOW);
  digitalWrite (B, LOW);
  digitalWrite (C, LOW);
  digitalWrite (D, LOW);
  digitalWrite (E, HIGH);
  digitalWrite (F, HIGH);
  digitalWrite (G, LOW);
  break;
case 4:
  digitalWrite (A, HIGH);
  digitalWrite (B, LOW);
  digitalWrite (C, LOW);
  digitalWrite (D, HIGH);
  digitalWrite (E, HIGH);
  digitalWrite (F, LOW);
  digitalWrite (G, LOW);
  break;
case 5:
  digitalWrite (A, LOW);
  digitalWrite (B, HIGH);
  digitalWrite (C, LOW);
  digitalWrite (D, LOW);
  digitalWrite (E, HIGH);
```

```
      digitalWrite (F, LOW);
      digitalWrite (G, LOW);
      break;
   case 6:
      digitalWrite (A, LOW);
      digitalWrite (B, HIGH);
      digitalWrite (C, LOW);
      digitalWrite (D, LOW);
      digitalWrite (E, LOW);
      digitalWrite (F, LOW);
      digitalWrite (G, LOW);
      break;
   case 7:
      digitalWrite (A, LOW);
      digitalWrite (B, LOW);
      digitalWrite (C, LOW);
      digitalWrite (D, HIGH);
      digitalWrite (E, HIGH);
      digitalWrite (F, HIGH);
      digitalWrite (G, HIGH);
      break;
   case 8:
      digitalWrite (A, LOW);
      digitalWrite (B, LOW);
      digitalWrite (C, LOW);
      digitalWrite (D, LOW);
      digitalWrite (E, LOW);
      digitalWrite (F, LOW);
      digitalWrite (G, LOW);
      break;
   case 9:
      digitalWrite (A, LOW);
      digitalWrite (B, LOW);
      digitalWrite (C, LOW);
      digitalWrite (D, HIGH);
      digitalWrite (E, HIGH);
      digitalWrite (F, LOW);
      digitalWrite (G, LOW);
      break;
   }
   delay (10);
} while (timer < 100);
resetSubroutine();
```

```
    maxCount = maxCount + 1;
    timer = 0;
  }      //exits this while loop when maxCount = 10
  trigLatch = 0;//resets trigger latch, makes ready for next trigger
  maxCount = 0;
  timer = 0;
 } //ends trigState
}//end loop
 void resetSubroutine() {
   digitalWrite (A, HIGH); //turns off segment A //resets
   digitalWrite (B, HIGH); //turns off segment B
   digitalWrite (C, HIGH); //turns off segment C
   digitalWrite (D, HIGH); //turns off segment D
   digitalWrite (E, HIGH); //turns off segment E
   digitalWrite (F, HIGH); //turns off segment F
   digitalWrite (G, HIGH); //turns off segment G
 }
```

Dice Game with Seven-Segment Display (No Driver IC)

The code in this project (Listing 8-10) again uses the Arduino to directly drive each LED segment of the display without the use of a decoder-driver IC. The drawback is that more Arduino ports are needed and the wiring complexity is increased. As before, in wiring the circuit on a breadboard, it will be helpful to refer to the MAN72 display illustration in Figure 8-8, and some of the circuity shown in the schematic of Figure 8-11. We will again be multiplexing two seven-segment displays. This project adds some special effects as the number is displayed. The code also activates a speaker to play two sets of tones: A set of higher frequency tones will signal a win, and a lower steady tone is emitted if there is a loss. See the schematic in Figure 8-7 about how to connect a speaker to the Arduino. A series resistor must be used to limit excessive speaker current, or damage could result.

Because this code signals wins and losses, we need to understand the rules of the game of craps. There are two sets of rules that are differentiated between the first in a series of rolls and that of subsequent rolls. If on the beginning first roll, or just after a win or a loss occurs, the winning numbers are 7 or 11. The losing numbers are 2, 3, or 12. If neither a win nor a loss occurs in the first roll, then the number rolled is termed the *point*. The player continues throwing the dice in hopes of rolling the point. If, however, a 7 appears before the value of the point is matched, then the player will lose the round.

Listing 8-10. Dice Game with a Seven-Segment Display and No IC

```
const int A = 1; //pin names
const int B = 13;
const int C = 10;
const int D = 8;
const int E = 7;
const int F = 2;
const int G = 11;
const int trans1 = 3;
const int trans2 = 4;
const int trigPin = 6;
const int buzzer = 5;
boolean trigState = HIGH; //variables
int x;
int rollCounter;
int roll;
int point;
boolean win;
int   randomNum1;
int   randomNum2;
int dicePiece;
int loopDisplay;
int multiplex;
void setup() {
  pinMode (A, OUTPUT);   //display segments
  pinMode (B, OUTPUT);
  pinMode (C, OUTPUT);
  pinMode (D, OUTPUT);
  pinMode (E, OUTPUT);
```

```
  pinMode (F, OUTPUT);
  pinMode (G, OUTPUT);
  pinMode (buzzer, OUTPUT);
  pinMode (trans1, OUTPUT); //display selection transistors
  pinMode (trans2, OUTPUT);
  pinMode (trigPin, INPUT_PULLUP);
  randomSeed (analogRead(5));
}
void loop() {
  trigState = digitalRead (trigPin);
  if (trigState == LOW) {
    resetSubroutine(); //calls the reset subroutine
    randomNum1 = random (1, 7); //generates the dice numbers
    randomNum2 = random (1, 7);
    rollCounter = rollCounter + 1;
    roll = randomNum1 + randomNum2;
    for (x = 0; x < 3; x++) { //used to go around a circle 3 times
      while (loopDisplay < 6) {//0 through 5 cases for circle segments
        switch (loopDisplay) {  //this section circles the dice
          case 0:
            digitalWrite (A, LOW); //turns on segment A
            while (multiplex < 5) {
              digitalWrite (trans1, HIGH); //turns on display 1
              digitalWrite (trans2, LOW); //turns off display 2
              delay (5);
              digitalWrite (trans1, LOW); //turns off display 1
              digitalWrite (trans2, HIGH); //turns on display 2
              delay (5);
              multiplex ++;
            }
            loopDisplay++;
            multiplex = 0;
            break;
          case 1:
            digitalWrite (B, LOW); //turns on segment B
            while (multiplex < 5) {
              digitalWrite (trans1, HIGH); //turns on display 1
              digitalWrite (trans2, LOW); //turns off display 2
              delay (5);
              digitalWrite (trans1, LOW); //turns off display 1
              digitalWrite (trans2, HIGH); //turns on display 2
              delay (5);
              multiplex ++;
            }
```

```
  loopDisplay++;
  multiplex = 0;
  break;
case 2:
  digitalWrite (C, LOW); //turns on segment C
  while (multiplex < 5) {
    digitalWrite (trans1, HIGH); //turns on display 1
    digitalWrite (trans2, LOW); //turns off display 2
    delay (5);
    digitalWrite (trans1, LOW); //turns off display 1
    digitalWrite (trans2, HIGH); //turns on display 2
    delay (5);
    multiplex ++;
  }
  loopDisplay++;
  multiplex = 0;
  break;
case 3:
  digitalWrite (D, LOW); //turns on segment D
  while (multiplex < 5) {
    digitalWrite (trans1, HIGH); //turns on display 1
    digitalWrite (trans2, LOW); //turns off display 2
    delay (5);
    digitalWrite (trans1, LOW); //turns off display 1
    digitalWrite (trans2, HIGH); //turns on display 2
    delay (5);
    multiplex ++;
  }
  loopDisplay++;
  multiplex = 0;
  break;
case 4:
  digitalWrite (E, LOW); //turns on segment E
  while (multiplex < 5) {
    digitalWrite (trans1, HIGH); //turns on display 1
    digitalWrite (trans2, LOW); //turns off display 2
    delay (5);
    digitalWrite (trans1, LOW); //turns off display 1
    digitalWrite (trans2, HIGH); //turns on display 2
    delay (5);
    multiplex ++;
  }
```

```
        loopDisplay++;
        multiplex = 0;
        break;
    case 5:
        digitalWrite (F, LOW); //turns on segment F
        while (multiplex < 5) {
            digitalWrite (trans1, HIGH); //turns on display 1
            digitalWrite (trans2, LOW); //turns off display 2
            delay (5);
            digitalWrite (trans1, LOW); //turns off display 1
            digitalWrite (trans2, HIGH); //turns on display 2
            delay (5);
            multiplex ++;
        }
        loopDisplay++;
        multiplex = 0;
        break;
    }
    resetSubroutine(); //calls the reset subroutine
}
        loopDisplay = 0;
        } //ends segment circle
    } //ends trigState
    if (x == 3) { //x is 3 after the segments have circled 3 times
        x = 0; //reusing the x counter
        if (rollCounter == 1) {
            switch (roll) {
                case 7: //win
                    tone (buzzer, 4000);
                    win = 1;
                    rollCounter = 0;
                    break;
                case 11:  //win
                    tone (buzzer, 4000);
                    win = 1;
                    rollCounter = 0;
                    break;
                case 2:  //lose
                    tone (buzzer, 600);
                    rollCounter = 0;
                    break;
                case 3:  //lose
                    tone (buzzer, 600);
                    rollCounter = 0;
                    break;
```

```
      case 12:  //lose
        tone (buzzer, 600);
        rollCounter = 0;
        break;

   }//end switch case
   point = roll; //does this if no case met
}//end if
if (rollCounter > 1) {
   if (roll == point) {  //win, hits point
     tone (buzzer, 4000);
     win = 1;
     rollCounter = 0;
   }
   if (roll == 7) {  //lose
     tone (buzzer, 600);
     rollCounter = 0;
   }
}//end if
do {
   ////DISPLAY SECTION////
   digitalWrite (trans1, HIGH); //turns on display 1
   dicePiece = randomNum1; //cases for first die
   displaySubroutine();
   delay(10);
   resetSubroutine();
   digitalWrite (trans2, HIGH); //turns on display 2
   dicePiece = randomNum2; //cases for second die
   displaySubroutine();
   delay(10);
   resetSubroutine();
   x++; //increment the counter
   if (win == 1) {
     switch (x) {  //pulses winning tone
       case 25:  //shuts off
         noTone(buzzer);
         break;
       case 50: //turns on
         tone (buzzer, 4000);
         break;
       case 75://shuts off
         noTone (buzzer);
         break;
```

```
        case 100:
          tone (buzzer, 4000);   //turns on
          break;
        case 125:
          noTone (buzzer); //shuts off
          break;
        case 150:
          tone (buzzer, 4000);   //turns on
          win = 0;
          break;
      }
    }
  } while (x < 200);     //end of do section loop
  noTone(buzzer);
  win = 0;
  } //end if for starting numbers display
  x = 0;
} //ends loop
void resetSubroutine() {
  digitalWrite (A, HIGH); //turns off segment A //resets
  digitalWrite (B, HIGH); //turns off segment B
  digitalWrite (C, HIGH); //turns off segment C
  digitalWrite (D, HIGH); //turns off segment D
  digitalWrite (E, HIGH); //turns off segment E
  digitalWrite (F, HIGH); //turns off segment F
  digitalWrite (G, HIGH); //turns off segment G
  digitalWrite (trans1, LOW); //turns off display 1
  digitalWrite (trans2, LOW); //turns off display 2
}
void displaySubroutine() {
  switch (dicePiece) {
    case 1:
      digitalWrite (A, HIGH); //segments low light up
      digitalWrite (B, LOW);
      digitalWrite (C, LOW);
      digitalWrite (D, HIGH);
      digitalWrite (E, HIGH);
      digitalWrite (F, HIGH);
      digitalWrite (G, HIGH);
      break;
    case 2:
      digitalWrite (A, LOW);
      digitalWrite (B, LOW);
      digitalWrite (C, HIGH);
```

```
      digitalWrite (D, LOW);
      digitalWrite (E, LOW);
      digitalWrite (F, HIGH);
      digitalWrite (G, LOW);
      break;
    case 3:
      digitalWrite (A, LOW);
      digitalWrite (B, LOW);
      digitalWrite (C, LOW);
      digitalWrite (D, LOW);
      digitalWrite (E, HIGH);
      digitalWrite (F, HIGH);
      digitalWrite (G, LOW);
      break;
    case 4:
      digitalWrite (A, HIGH);
      digitalWrite (B, LOW);
      digitalWrite (C, LOW);
      digitalWrite (D, HIGH);
      digitalWrite (E, HIGH);
      digitalWrite (F, LOW);
      digitalWrite (G, LOW);
      break;
    case 5:
      digitalWrite (A, LOW);
      digitalWrite (B, HIGH);
      digitalWrite (C, LOW);
      digitalWrite (D, LOW);
      digitalWrite (E, HIGH);
      digitalWrite (F, LOW);
      digitalWrite (G, LOW);
      break;

    case 6:
      digitalWrite (A, LOW);
      digitalWrite (B, HIGH);
      digitalWrite (C, LOW);
      digitalWrite (D, LOW);
      digitalWrite (E, LOW);
      digitalWrite (F, LOW);
      digitalWrite (G, LOW);
      break;
  }
}
```

Electronic Dice Game with LEDs

This project (Listing 8-11) uses a series of two sets of seven LEDs arranged in the pattern of the dots on two dice pieces (Figure 8-12). The positive anodes of each LED are connected in parallel to its companion display to produce a data bus similar to the one in Figure 8-11. To add a light that signals a win, connect an LED from ground through a 220 Ohm resistor to pin A2. Connect a buzzer or speaker to signal a win or a loss, by connecting through a 120 Ohm resistor to pin 11, similar to Figure 8-7. We are using the analog pins as digital outputs. Pins A0 through A5 are very versatile by having the ability to be used for either analog or digital I/O. The trigger pin for the game is pin 8. It can be momentarily grounded by a player to start a new roll or can be kept grounded to run the game in demonstration mode.

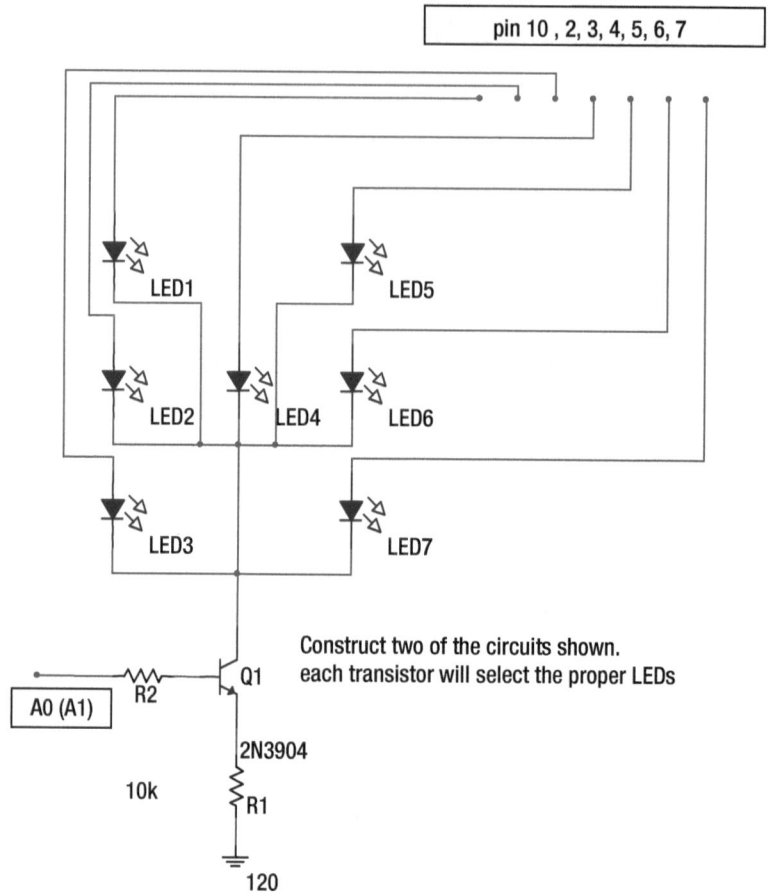

Figure 8-12. *LEDs configured for dice game*

One of the circuits is shown in Figure 8-13. The best way to connect complex circuits like this is to start at whatever point you like, and then connect each spot wire by wire. A wise man once advised against looking at the overall project and instead just concentrating on each small section at any given time. Some people like to check off each connection on the schematic with a pencil as they are made on a breadboard. It is a translation to go from an orderly schematic to the actual world and we get better with practice. It is important to be focused and not be in a hurry, and also to double check as each connection is made.

198

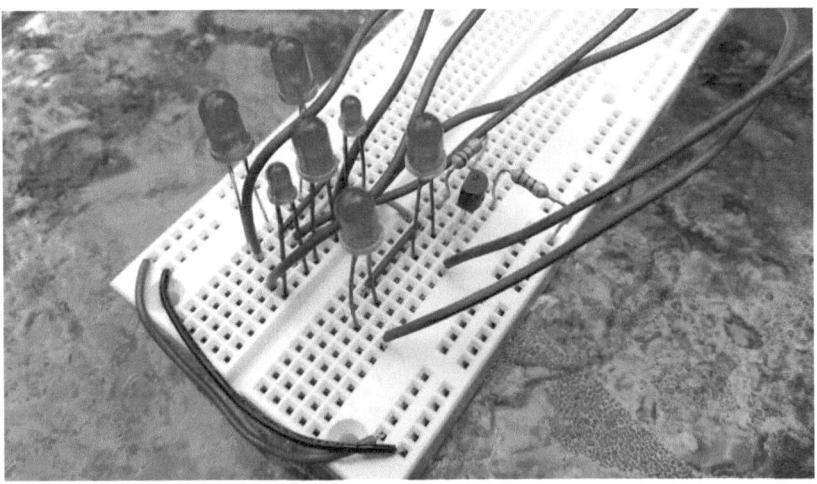

Figure 8-13. *One of the LED dice circuits wired on a breadboard*

DICE GAME RULES RECAP

Winning: 7 or 11 on first pass lights win LED.

Losing: 2, 3, or 12 on first pass sounds buzzer.

If no win or loss on the first pass, play for the point with the following conditions:

A point match lights the LED, but hitting 7 is a bust and sounds the buzzer.

All numbers are displayed on the pair of seven LEDs, each arranged in a dice pattern.

Listing 8-11. Electronic Dice Game with LEDs

```
int randomNum1;    //LED dice game
int randomNum2;
const int trigger = 8;
const int winLED = A2;
const int buzzer = 11;
int trigState;
int time;
int roll;
int point;
int rollCounter;
int win;
void setup(){
  pinMode (trigger, INPUT_PULLUP); //throw the dice switch
  pinMode (10, OUTPUT); //LED 1
  pinMode (2, OUTPUT); //LED 2
  pinMode (3, OUTPUT); //LED 3
  pinMode (4, OUTPUT); //LED 4
  pinMode (5, OUTPUT); //LED 5
  pinMode (6, OUTPUT); //LED 6
  pinMode (7, OUTPUT); //LED 7
  pinMode (A0, OUTPUT); // die driver 1
  pinMode (A1, OUTPUT); //die driver 2
  pinMode (winLED, OUTPUT);
  pinMode (buzzer, OUTPUT);
  randomSeed(analogRead(5));//gets random number from noise on pin A5
}
void loop() {//looks for button press and displays two random numbers
  trigState = digitalRead (trigger); //pin 8
  while (trigState == 0){  //can run program in display mode
    //numbers 1 through 6 quickly flash for a nice visual effect
    digitalWrite(A0, HIGH);//selects dice one
    digitalWrite(A1, LOW);
    digitalWrite (4, HIGH);    //displays a 1
    delay (90);
    digitalWrite(A0, LOW);
    digitalWrite(A1, HIGH);    //selects dice two
    delay (90);
    digitalWrite (4, LOW);     //shuts off 1
```

```
digitalWrite(A0, HIGH);      //selects dice one
digitalWrite(A1, LOW);
digitalWrite (10, HIGH);     //displays a 2
digitalWrite (7, HIGH);
delay (90);
digitalWrite(A0, LOW);
digitalWrite(A1, HIGH);     //selects dice two
delay (90);
digitalWrite (10, LOW);     //shuts off 2
digitalWrite (7, LOW);

digitalWrite(A0, HIGH);      //selects dice one
digitalWrite(A1, LOW);
digitalWrite (10, HIGH);     //displays a 3
digitalWrite (4, HIGH);
digitalWrite (7, HIGH);
delay (90);
digitalWrite(A0, LOW);
digitalWrite(A1, HIGH);     //selects dice two
delay (90);
digitalWrite (10, LOW);     //shuts off 3
digitalWrite (4, LOW);
digitalWrite (7, LOW);

digitalWrite(A0, HIGH);      //selects dice one
digitalWrite(A1, LOW);
digitalWrite (10, HIGH);     //displays a 4
digitalWrite (3, HIGH);
digitalWrite (5, HIGH);
digitalWrite (7, HIGH);
delay (90);
digitalWrite(A0, LOW);
digitalWrite(A1, HIGH);     //selects dice two
delay (90);
digitalWrite (10, LOW);     //shuts off 4
digitalWrite (3, LOW);
digitalWrite (5, LOW);
digitalWrite (7, LOW);

digitalWrite(A0, HIGH);      //selects dice one
digitalWrite(A1, LOW);
digitalWrite (10, HIGH);     //displays a 5
digitalWrite (3, HIGH);
```

201

```
digitalWrite (4, HIGH);
digitalWrite (5, HIGH);
digitalWrite (7, HIGH);
delay (90);
digitalWrite(A0, LOW);
digitalWrite(A1, HIGH);    //selects dice two
delay (90);
digitalWrite (10, LOW);     //shuts off 5
digitalWrite (3, LOW);
digitalWrite (4, LOW);
digitalWrite (5, LOW);
digitalWrite (7, LOW);

digitalWrite(A0, HIGH);     //selects dice one
digitalWrite(A1, LOW);
digitalWrite (10, HIGH);     //displays a 6
digitalWrite (2, HIGH);
digitalWrite (3, HIGH);
digitalWrite (5, HIGH);
digitalWrite (6, HIGH);
digitalWrite (7, HIGH);
delay (90);
digitalWrite(A0, LOW);
digitalWrite(A1, HIGH);    //selects dice two
delay (90);
digitalWrite (10, LOW);     //shuts off 6
digitalWrite (2, LOW);
digitalWrite (3, LOW);
digitalWrite (5, LOW);
digitalWrite (6, LOW);
digitalWrite (7, LOW);
//start game
rollCounter = rollCounter +1; //rollCounter keeps track of rolls
randomNum1 = random(1, 7);
randomNum2 = random(1, 7);
roll = randomNum1 + randomNum2; //dice total number
if( roll == 2 && rollCounter == 1){ //tests for snake eyes
                                 //on first pass
   tone(buzzer, 400);
   rollCounter = 0;
}
```

```
if (roll == 3 && rollCounter == 1){ //tests for a loss craps
                                    //on first pass
   tone(buzzer, 400);
   rollCounter = 0;
}
if (roll == 12 && rollCounter == 1) { //busted with 12
                                      //on first pass
   tone(buzzer, 400);
   rollCounter = 0;
}
if (roll == 7 && rollCounter >1) { //after first pass,
                                   //playing for point but get 7
   tone(buzzer, 400);
   rollCounter = 0;
}
if (roll == 7 && rollCounter == 1) { //win, 7 on first pass
   digitalWrite(winLED, HIGH);
   tone(buzzer, 4000);
   win = 1;
   rollCounter = 0;
}
if  (roll == 11 && rollCounter == 1){//win, 11 on first pass
   digitalWrite(winLED, HIGH);
   tone(buzzer, 4000);
   win = 1;
   rollCounter = 0;
}
if (roll == point && rollCounter > 1){//win, hit point
   digitalWrite(winLED, HIGH);
   tone(buzzer, 4000);
   win = 1;
   rollCounter = 0;
}
do {
   time = time + 1;  //the do loops around 249 times,
   digitalWrite(A0, HIGH);   //selects dice one
   digitalWrite(A1, LOW);    //deselects dice two
   switch (randomNum1) {     //picks the first dice LEDs to light
```

```
case 1:
  digitalWrite (4, HIGH);
  break;
case 2:
  digitalWrite (10, HIGH);
  digitalWrite (7, HIGH);
  break;
case 3:
  digitalWrite (10, HIGH);
  digitalWrite (4, HIGH);
  digitalWrite (7, HIGH);
  break;
case 4:
  digitalWrite (10, HIGH);
  digitalWrite (3, HIGH);
  digitalWrite (5, HIGH);
  digitalWrite (7, HIGH);
  break;
case 5:
  digitalWrite (10, HIGH);
  digitalWrite (3, HIGH);
  digitalWrite (4, HIGH);
  digitalWrite (5, HIGH);
  digitalWrite (7, HIGH);
  break;
case 6:
  digitalWrite (10, HIGH);
  digitalWrite (2, HIGH);
  digitalWrite (3, HIGH);
  digitalWrite (5, HIGH);
  digitalWrite (6, HIGH);
  digitalWrite (7, HIGH);
  break;
}
delay (10);
digitalWrite (10, LOW);    //shuts off the first dice LEDs
digitalWrite (2, LOW);
digitalWrite (3, LOW);
digitalWrite (4, LOW);
digitalWrite (5, LOW);
digitalWrite (6, LOW);
digitalWrite (7, LOW);
```

```
  digitalWrite(A0, LOW);          //deselects dice one
  digitalWrite(A1, HIGH);         //selects dice two
  switch (randomNum2) {    //picks the second dice LEDs to light
case 1:
  digitalWrite (4, HIGH);
  break;
case 2:
  digitalWrite (10, HIGH);
  digitalWrite (7, HIGH);
  break;
case 3:
  digitalWrite (10, HIGH);
  digitalWrite (4, HIGH);
  digitalWrite (7, HIGH);
  break;
case 4:
  digitalWrite (10, HIGH);
  digitalWrite (3, HIGH);
  digitalWrite (5, HIGH);
  digitalWrite (7, HIGH);
  break;
case 5:
  digitalWrite (10, HIGH);
  digitalWrite (3, HIGH);
  digitalWrite (4, HIGH);
  digitalWrite (5, HIGH);
  digitalWrite (7, HIGH);
  break;
case 6:
  digitalWrite (10, HIGH);
  digitalWrite (2, HIGH);
  digitalWrite (3, HIGH);
  digitalWrite (5, HIGH);
  digitalWrite (6, HIGH);
  digitalWrite (7, HIGH);
  break;
}
delay (10);
digitalWrite (10, LOW);         //shuts off the second dice LEDs
digitalWrite (2, LOW);
digitalWrite (3, LOW);
digitalWrite (4, LOW);
```

```
    digitalWrite (5, LOW);
    digitalWrite (6, LOW);
    digitalWrite (7, LOW);
    if (win == 1 && time == 10){ //winning buzzer beeps
      noTone(buzzer);
    }
    if (win == 1 && time == 20){
      tone(buzzer, 4000);
    }
    if (win == 1 && time == 30){
      noTone(buzzer);
    }
    if (win == 1 && time == 40){
      tone(buzzer, 4000);
    }
    if (win == 1 && time == 50){
      noTone(buzzer);
    }
    if (time > 70 && win != 1) {  //short buzzer sound for loss
      noTone(buzzer);
    }
  }
  while (time < 250);//creates loop for both dice LEDs to be visible

  if (rollCounter == 1) { //if the pass went without win or loss
     point = roll;       //loads the point in memory
  }
  trigState = 1;  //resets trigger flag for program activation
  time = 0;       //resets loop timer for LED display
  digitalWrite(A0, LOW);
  digitalWrite(A1, LOW);
  digitalWrite(winLED, LOW);
  win = 0;
  }
}
```

Review Questions

1. The output pulses from a 555 timer are

   a. square waves.

   b. sine waves.

   c. circular waves.

   d. longitudinal waves.

2. The gap running down the center of a breadboard work area

   a. provides the voltage.

   b. separates the two work area sides.

   c. allows for adequate cooling.

   d. is a place for the convenient placement of wire.

3. A piezoelectric element exhibits reciprocity that most closely equates to which of the following statements?

   a. Letter A comes before B.

   b. The transistor was used to replace vacuum tubes.

   c. A single radio antenna can be used for transmission and reception.

   d. Six plus six equals twelve, not a baker's dozen.

4. A very rapidly flashing LED might look

   a. dimmer than if it were on continuously.

   b. brighter than if it were on continuously.

   c. slightly a different color.

   d. slightly a different hue.

5. Displays to produce the numbers 0 through 9 have
    _____segments.

    a.   two

    b.   seven

    c.   eight

    d.   nine

6. _____binary digits are required to produce a
    decimal output of 0 to 9.

    a.   Two

    b.   Three

    c.   Four

    d.   Ten

7. Which of these is the code to output either a high or
    low from the Arduino?

    a.   `digitalRead`

    b.   `digitalWrite`

    c.   `pulseOut`

    d.   `count`

8. LEDs have a positive and negative side.
    (True/False)

9. The best way to connect circuits on a breadboard is

    a.   as quickly as possible.

    b.   one wire at a time.

  c. positives all followed by negatives.

  d. with power on.

10. Multiple true conditions can be tested with which operator to see if two or more conditions exist simultaneously?

  a. OR (||)

  b. AND (&&)

  c. If

  d. While

Project 8

As was mentioned in the chapter in describing the operation of a piezoelectric sounder, electrical energy is converted to mechanical energy and the ultimate conversion is into heat. Describe the process.

CHAPTER 9

More Elaborate Projects

Coding a More Functional Poker Game

In Chapter 6 we experimented with coding the game of poker and were able to put together a very simplistic version of the game. The advanced program in this chapter builds on the earlier code for generating nonduplicate numbers to represent the cards in a playing deck. It deals five random cards to the player and five to the dealer (refer to Table 6-1 for card value translations). The code in Listing 9-1 goes on to identify both hands by using the rules of poker. At first glance, it looks like a tremendous amount of code, but because the player and dealer sections are similar, much of it can be copied and pasted with only minor alterations. It is left to the reader to generate the additional code needed to identify a winner if desired. My intent is to only show how a computer program can act in accordance with a set of rules. You might also wish to rework the code using functions (subroutines) to eliminate some repetition and provide for better overall organization.

© Bob Dukish 2018
B. Dukish, *Coding the Arduino*, https://doi.org/10.1007/978-1-4842-3510-2_9

Listing 9-1. An Expanded Poker Game

```
int ArrayOne[72]; //produces ten non-duplicate numbers efficiently
int ArrayMirror[72]; //progressively adds to the non-dupe array
int ArrayDupes[72]; //used to deal two hands of 5 card poker
int ArrayDeal [52];
byte ArrayPlayerCard [5];
byte ArrayPlayerSuit [5];
byte ArrayDealerCard [10];
byte ArrayDealerSuit [10];
byte suit;
byte card;
byte x;
byte i;
byte j;
byte d;
byte match;
byte cardMatched;
const byte trigger = 7;
boolean trigState;
boolean goToReset;
boolean pair;
boolean twoPair;
boolean threeOfKind;
boolean fourOfKind;
boolean fullHouse;
byte matchedCard;
byte matchedCardTwo;
byte hearts;
byte spades;
byte diamonds;
byte clubs;
byte largestPlayerCard;
byte largestDealerCard;
byte smallestPlayerCard;
byte smallestDealerCard;
byte straight;
byte distance;
boolean suitFlush;
byte royalFlush;
byte playerSuit;
byte dealerSuit;
boolean playerAce;
boolean dealerAce;

void setup() {
  pinMode (trigger, INPUT_PULLUP);
```

```
  randomSeed (analogRead(5));
  Serial.begin (9600);
  Serial.println ("Ground pin 7 to start");
  Serial.println (" ");
}
void loop(){
  trigState = digitalRead (trigger);
  delay (100);
  while (trigState == LOW){//this section sets up numbers,makes copy
    Serial.println (" ");
    for (i = 0; i < 72; i++){
      ArrayOne [i] = random (1, 74);
      ArrayMirror [j] = ArrayOne [i];
      j++;
    }
reset:                     //restarts here to clear duplicate numbers
    goToReset = false;

    for (d = 0; d < 72; d++){     //zeros the dupe array
      ArrayDupes [d] = 0;
    }
    for (i = 0; i < 72; i++){
      for (j = 0; j < 72; j++){
        if ( (ArrayOne [i] == ArrayMirror [j]) && (i > j) && (i != 0
)){
          //tests for dupes
          ArrayDupes [i] = ArrayOne [i]; // gives an array of dupes,
        }                              // and zeros for no dupes
      }
    }
    i = 0;
    for (d = 0; d < 72; d++){//if ArrayDupes number is dupe, pick new
      if (ArrayDupes [d] != 0){
        ArrayOne [d] = random (1, 74);
        goToReset = true;
      }
    }
    j =0;
    for (i = 0; i < 72; i++){             //loads mirror array
      ArrayMirror [j] = ArrayOne [i];
      j++;
    }
```

```
    if (goToReset == true){
      goto reset;// since there were dupes, go back to run test again
    }
    for (i = 0; i < 52; i++){//Now no dupes, get good random  numbers

      if ((ArrayOne [i] < 14) || ((ArrayOne [i] >20) && (ArrayOne
[i] < 34))  ||
        ((ArrayOne [i] > 40) && (ArrayOne [i] < 53))|| (ArrayOne [i]
> 60)){
        ArrayDeal [x] = ArrayOne [i];
        x++;
      }
    }
    //***************testing section***************
    //section below if added gives player royal flush and dealer 1
      //pair for testing
    //ArrayDeal[0] = 1;  // this section tests player possibilities
    //ArrayDeal[1] = 13; //remove to run game
    //ArrayDeal[2] = 12; //install to test
    //ArrayDeal[3] = 11;
    //ArrayDeal[4] = 10;

    //ArrayDeal[5] = 23;  // this section tests dealer possibilities
    //ArrayDeal[6] = 43; //remove to run game
    //ArrayDeal[7] = 61; //install to test
    //ArrayDeal[8] = 5;
    //ArrayDeal[9] = 2;
    //*********end of testing section************
    Serial.println (" ");
    for (x = 0; x < 10; x++){     //converting array to cards
      if (ArrayDeal [x] > 60){
        suit = 3;
        card = ArrayDeal [x] - 60;
      }
      else if (ArrayDeal [x] > 40){
        suit = 2;
        card = ArrayDeal [x] - 40;
      }
      else if (ArrayDeal [x] > 20){
        suit = 1;
        card = ArrayDeal [x] - 20;
      }
      else if (ArrayDeal [x] < 14){
```

```
  suit = 0;
  card = ArrayDeal [x];
  //card = card
}
Serial.println (" ");
if (x == 0) {
  Serial.println ("Player's Cards:");
  Serial.println (" ");
}
if (x == 5){
  Serial.println (" ");
  Serial.println ("****************");
  Serial.println (" ");
  Serial.println ("Dealer's Cards:");
  delay (1000);
  Serial.println (" ");
}
if (card == 1){
  Serial.print ("The Ace");//converting cards larger than 10
}
else if (card == 11){
  Serial.print ("Jack");
}
else if (card == 12){
  Serial.print ("Queen");
}
else if (card == 13){
  Serial.print ("King");
}
else{
  Serial.print (card); // for 10 to two, shows cards directly
}
Serial.print (" of ");
switch (suit) {              //shows suit
case 3:
  Serial.print ("Hearts");
  hearts = 1;
  break;
case 2:
  Serial.print ("Spades");
  spades = 1;
  break;
```

```
case 1:
  Serial.print ("Diamonds");
  diamonds = 1;
  break;
case 0:
  Serial.print ("Clubs");
  clubs = 1;
  break;
}
//loads arrays for outcome
if(x < 5){
  if (card == 1){ //check for player ace for straight
    playerAce = true;
  }
  ArrayPlayerCard[x] = card;
  ArrayPlayerSuit[x] = suit;
  playerSuit = clubs + diamonds + spades + hearts;
}
if (x == 4){//zeros suit after player, checks for dealer flush
  clubs = 0;
  diamonds = 0;
  spades = 0;
  hearts = 0;
}
if ((x > 4) && ( x < 10)){
  if (card == 1){
    dealerAce = true; //check for dealer ace for straight
  }
  ArrayDealerCard[x] = card;
  ArrayDealerSuit[x] = suit;
  dealerSuit = clubs + diamonds + spades + hearts;
}
}
Serial.println (" ");
smallestPlayerCard = 13;//set up for finding smallest card before
                        //entering loop
for (x = 0; x < 5; x++){          //player outcome follows:
  for (i = 0; i < 5; i++){
    if((ArrayPlayerCard[i] == ArrayPlayerCard[x]) && (i > x) && (i
!= 0)){                       //if a match enter loop
      match = match + 1;
      if (match == 1){
        matchedCard = ArrayPlayerCard[i]; //one pair
        pair = true;
      }
```

```
      if((match == 2) && (matchedCard != ArrayPlayerCard[i])){
        twoPair = true;
      }
      if((match == 3) && (matchedCard == ArrayPlayerCard[i])){
        threeOfKind = true;
      }
      else if(match == 4){
        fullHouse = true;
      }
      if((match == 4) && (matchedCard == ArrayPlayerCard[i]) &&
        (threeOfKind == true)){
        fourOfKind = true;
      }
    }
    if ((ArrayPlayerCard[x] > ArrayPlayerCard[i]) &&
(ArrayPlayerCard[x] > largestPlayerCard)){//finds player highest
      largestPlayerCard = ArrayPlayerCard[x];
    }
    if ((ArrayPlayerCard[x] < ArrayPlayerCard[i]) &&
(ArrayPlayerCard[x] < smallestPlayerCard) && (ArrayPlayerCard[x] !=
1)){  //finds smallest
      smallestPlayerCard = ArrayPlayerCard[x];
    }
  }
} // end of player for loop
distance = largestPlayerCard - smallestPlayerCard;//checks for
                                                  //straight
if ((distance  == 4) && (match == 0) && (playerAce != true)){
  straight = true;
}
if (((distance == 3) && (largestPlayerCard == 5) && (match == 0)
&& (playerAce == true)) || //check for high/low ace straight
  ((distance == 3) && (largestPlayerCard == 13)&& (match == 0) &&
(playerAce == true))){
  straight = true;
}
if (playerSuit == 1){
  suitFlush = true;
}
```

```
    if ((straight == true) && (suitFlush == true) &&
(largestPlayerCard == 13) && (playerAce == true)){
      royalFlush = true;
    }
    delay (1000);
    Serial.println (" ");
    Serial.println (" ");
    Serial.print ("Player's hand is ");       //condition testing
    if (royalFlush == true){
      Serial.print ("royal flush ");
    }
    else if (fourOfKind == true){
      Serial.print ("four of a kind  ");
    }
    else if ((suitFlush == true ) && (straight == true)){
      Serial.print ("a straight flush  ");
    }
    else if (fullHouse == true){
      Serial.print ("full house  ");
    }
    else if (suitFlush == true){
      Serial.print ("a flush  ");
    }
    else if (straight == true) {
      Serial.print ("a straight  ");
    }
    else if (threeOfKind == true){
      Serial.print ("three of a kind  ");
    }
    else if (twoPair == true){
      Serial.print ("two pair  ");
    }
    else if (pair == true){
      Serial.print ("one pair  ");
    }
    else{
      Serial.print("high card ");
    }
    match = 0;
    pair = false;
    twoPair = false;
    threeOfKind = false;
    fourOfKind = false;
    fullHouse = false;
    matchedCard = 0;
    matchedCardTwo = 0;
    straight = false;
```

```
    suitFlush = false;
    royalFlush = false;
    distance = 0;
    //dealer outcome follows:
    smallestDealerCard = 13; //set up for finding smallest card before
                             //entering loop
    for (x = 5; x < 10; x++){
      for (i = 5; i < 10; i++){
        if((ArrayDealerCard[i] == ArrayDealerCard[x])&&(i > x)&&(i
!= 0)){                                     //if a match, enter loop
          match = match + 1;
          if (match == 1){
            matchedCard = ArrayDealerCard[i];
            pair = true;
          }
          if((match == 2) &&(matchedCard != ArrayDealerCard[i])){
            twoPair = true;
          }
          if((match == 3) && (matchedCard == ArrayDealerCard[i])){
            threeOfKind = true;
          }
          else if(match == 4){
            fullHouse = true;
          }
          if((match == 4) && (matchedCard == ArrayDealerCard[i]) &&
      (threeOfKind == true)){
            fourOfKind = true;
          }
        }
        if ((ArrayDealerCard[x] > ArrayDealerCard[i]) &&
(ArrayDealerCard[x] > largestDealerCard)){  //finds the high card
          largestDealerCard = ArrayDealerCard[x];
        }

        if ((ArrayDealerCard[x] < ArrayDealerCard[i]) &&
(ArrayDealerCard[x] < smallestDealerCard)`&& (ArrayDealerCard[x] !=
1)){  //finds dealer small card
          smallestDealerCard = ArrayDealerCard[x];
        }
      }
    }                       // end of dealer for loop
    distance = (largestDealerCard - smallestDealerCard); //checks for
                                                //straight
```

219

```
    if ((distance  == 4) && (match == 0) && (dealerAce != true)){
      straight = true;
    }
    if (((distance == 3) && (largestDealerCard == 5) && (match == 0)
&& (dealerAce == true)) || //check for high/low ace straight
      ((distance == 3) && (largestDealerCard == 13) && (match == 0) &&
(dealerAce == true))){
      straight = true;
    }
    if (dealerSuit == 1){
      suitFlush = true;
    }
    if ((straight == true) && (suitFlush == true) &&
(largestDealerCard == 13) && (dealerAce == true)){
      royalFlush = true;
    }
    delay (1000);
    Serial.println (" ");
    Serial.println (" ");
    Serial.print ("Dealer's hand is ");       //condition testing
    if (royalFlush == true){
      Serial.print ("Royal Flush ");
    }
    else if (fourOfKind == true){
      Serial.print ("four of a kind  ");
    }
    else if ((suitFlush == true ) && (straight == true)){
      Serial.print ("a straight flush  ");
    }
    else if (fullHouse == true){
      Serial.print ("full house  ");
    }
    else if (suitFlush == true){
      Serial.print ("a flush  ");
    }
    else if (straight == true) {
      Serial.print ("a straight  ");
    }
    else if (threeOfKind == true){
      Serial.print ("three of a kind  ");
    }
    else if (twoPair == true){
      Serial.print ("two pair  ");
    }
```

```
    else if (pair == true){
      Serial.print ("one pair  ");
    }
    else{
      Serial.print("high card ");
    }
    Serial.println (" ");
    x = 0;
    i = 0;
    j = 0;
    d = 0;
    match = 0;
    pair = 0;
    twoPair = 0;
    threeOfKind = 0;
    fourOfKind = 0;
    fullHouse = 0;
    matchedCard = 0;
    matchedCardTwo = 0;
    largestPlayerCard = 0;
    largestDealerCard = 0;
    smallestPlayerCard = 0;
    smallestDealerCard = 0;
    straight = 0;
    playerSuit = 0;
    dealerSuit = 0;
    suitFlush = 0;
    royalFlush = 0;
    distance = 0;
    playerAce = 0;
    dealerAce = 0;
    for (i = 0; i < 52; i++){
      ArrayOne[i] = 0;
    }
    for (i = 0; i < 10; i++){    //zeros at end of the game
      ArrayPlayerCard[i];
      ArrayPlayerSuit[i];
      ArrayDealerCard[i];
      ArrayDealerSuit[i];
    }
    Serial.println (" ");
    Serial.println
("_____");
    Serial.println (" ");
    trigState = HIGH;
  }
}
```

Coding a More Functional Game of 21

In Chapter 7, we made two attempts to code the game of 21 (blackjack). This advanced project, the code for which is shown in Listing 9-2, significantly expands on the first and second versions of the game. It uses pin 7 as before to begin the game, but now includes two hardware interrupts that allow the player to make the decision of adding the points from additional cards to surpass the dealer's points without going over 21. Many of the actual game rules are incorporated into the code that follows, but it is left to the reader to finish the game so that when two equal cards are dealt to the player, he or she has the option to play two distinct hands, which is called a split. Also, note that the nonduplication that was coded into the last project is neglected to simulate a dealer using multiple card decks, but can be incorporated into the game, if desired.

Listing 9-2. Expanded Functionality for a Game of 21

```
/*BlackJack (21) game
 Written 6/15/16 as a learning tool to aid students
 in exploring program development.*/
byte hello;
byte randomNum1;
byte randomNum2;
byte randomNum3;
byte randomNum4;
byte randomNum5;
byte randomNum6;
boolean win;
boolean lose;
byte win_lose;
byte hits;
byte hitCounter;
byte dealer;
byte player;
byte playerSticks;
byte dealerSticks;
boolean trigState;
boolean trigLatch;
int timed;
```

```
boolean ace = false;
byte ace10;
byte ace1;
int aceLatch;
const byte trigger1 = 7;
const byte triggerYes = 2; //pin 2 is hardware interrupt 0
const byte triggerNo = 3; //pin 3 is hardware interrupt 1
void setup(){
  pinMode (trigger1, INPUT_PULLUP);
  pinMode (triggerYes, INPUT_PULLUP);
  pinMode (triggerNo, INPUT_PULLUP);
  Serial.begin(9600);
  randomSeed(analogRead(5));
  attachInterrupt (0, yes_ISR, LOW);
  attachInterrupt (1, no_ISR, LOW);
}

void loop(){
  if (hello == 0){      //messages outside of game loop
    delay (500);
    Serial.println (" ");
    Serial.println ("Welcome to The Game of Twenty One");
    Serial.println (" ");
    delay (1000);
    Serial.println ("Sorry, no splitting cards");
    Serial.println (" ");
    delay (1500);
    Serial.println ("Ground Pin 7, to Start the Game");
    Serial.println (" ");
    hello = 1;
  }
  if (hello == 2){
    delay (500);
    Serial.println (" ");
    Serial.print ( "To Play Again");
    Serial.print (" ");
    delay (1000);
    Serial.println ("Ground Pin 7");
    Serial.println (" ");
    hello = 3;
  }
```

```
if (hello == 5){
  delay (500);
  Serial.println (" ");
  Serial.print ( "Remember, casinos aren't in business to lose!");
  delay (1000);
  Serial.print (" ");
  Serial.println (" ");
  hello = 6;
}
if (hello == 8){
  delay (500);
  Serial.println (" ");
  Serial.print ( "Nothing is up my sleeve!");
  delay (1000);
  Serial.print (" ");
  Serial.println (" ");
  hello = 9;
}
randomNum5 = 0;
randomNum6 = 0;
win = 0;
lose =0;
win_lose = 0;
hitCounter = 0;
hits = 0;
dealer = 0;
player = 0;
playerSticks = 0;
dealerSticks = 0;
trigState = 0;
trigLatch = 0;
timed = 0;
ace = false;
ace10 = 0;
ace1 = 0;
aceLatch = 0;
trigState = digitalRead(trigger1); //game start
if (trigState == LOW){
  trigLatch = 1;
  hello++;
}
```

```
while (trigLatch == 1){
  delay (500);
  randomNum1 = random (2, 15);
  Serial.println (" ");
  Serial.println ("*********");
  Serial.println (" ");
  Serial.print ("Your First Card is "); //your first card:
  switch (randomNum1){
  case 11:
    Serial.print ("an Ace, worth ");
    ace = true;
    break;
  case 12:
    Serial.print ("a Jack, worth ");
    break;
  case 13:
    Serial.print ("a Queen, worth ");
    break;
  case 14:
    Serial.print ("a King, worth ");
    break;
  }
  if (randomNum1 > 11){
    randomNum1 = 10;
  }
  Serial.print (randomNum1);
  delay (1500);
  // your first card, end code:
  //****************
                // your second card, begin code:
  randomNum3 = random (2, 15);
  Serial.print (",     Your Second Card is ");
  switch (randomNum3){
  case 11:
    Serial.print ("an Ace, worth ");
    ace = true;
    break;
  case 12:
    Serial.print ("a Jack, worth ");
    break;
  case 13:
    Serial.print ("a Queen, worth ");
    break;
```

```
  case 14:
    Serial.print ("a King, worth ");
    break;
  }
  if (randomNum3 > 11){
    randomNum3 = 10;
  } //may remove to split
//not allow a player bust with 2 aces = 22, makes the second ace = 1
  if ((randomNum1 == 11) && (randomNum3 == 11)){
    randomNum3 = 1;
  }
  Serial.print (randomNum3);
  Serial.println (" ");
  Serial.println (" ");           // your second card, end code:
  delay (3000);
  //**********************
 // Dealer's first card, begin code:
  randomNum2 = random (2, 15);
  Serial.print ("The Dealer's First Card is ");
  switch (randomNum2){
  case 11:
    Serial.print ("an Ace, worth ");
    break;
  case 12:
    Serial.print ("a Jack, worth ");
    break;
  case 13:
    Serial.print ("a Queen, worth ");
    break;
  case 14:
    Serial.print ("a King, worth ");
    break;
  }
  if (randomNum2 > 11){
    randomNum2 = 10;
  }
  Serial.print (randomNum2);
  delay (2000);
  //Dealer's first card , end code
  //**********************
 // dealer's second card, begin code:
  randomNum4 = random (2, 15);
  Serial.print (",     The Dealer's Second Card is ");
```

```
switch (randomNum4){
case 11:
  Serial.print ("an Ace, worth ");
  break;
case 12:
  Serial.print ("a Jack, worth ");
  break;
case 13:
  Serial.print ("a Queen, worth ");
  break;
case 14:
  Serial.print ("a King, worth ");
  break;
}
if (randomNum4 > 11){
  randomNum4 = 10;
}
if ((randomNum2 == 11) && (randomNum4 == 11)){
  randomNum4 = 1;      //no dealer bust with 2 aces = 22,
}                      //makes the second ace = 1
Serial.print (randomNum4);
delay (3500);
Serial.println (" ");
// dealer's second card, end code:
//*************************
dealer = (randomNum2 + randomNum4);
player = (randomNum1 + randomNum3);
Serial.println (" ");
Serial.println (" ");
Serial.print ("Your Total is ");
Serial.println (player);
Serial.println (" ");
delay (1000);
Serial.print ("The Dealer's Total is ");
Serial.println (dealer);
delay (1000);
Serial.println (" ");
if (dealer == 21){ //dealer gets blackjack, you lose
  lose = 1;
  win_lose = 1;
}
if (dealer > 16) {
  Serial.println (" ");
```

```
      Serial.print ("*** Dealer Sticks at ");
      Serial.print (dealer);
      Serial.println (" ***");
      dealerSticks = 1;
   }


   if (ace == true){          //player selects Ace for 1 or 10 points
      Serial.println (" ");
      Serial.println (" ");
      Serial.print("??? Want the Ace to be 1 or 11 points ???");
      Serial.println (" ***(pin 2 for 11 points, pin 3 for 1 point)
***");
      timed = 0;
      while (timed < 6000){
        delay (10);
        if (timed == 3000){    //nag
          Serial.println (" ");
          Serial.println ("Ace 1 or 11 points ?");
          Serial.println (" ");
          Serial.println ("If no decision in 30 seconds, it remains 11
points");
          Serial.println (" ");
        }
        if (timed == 4500){
          Serial.println (" ");
          Serial.println ("Ace 1 or 11 points ?");
          Serial.println (" ");
        }
        timed++;
      }
      delay (10);
      switch (aceLatch){
      case 10:
        player = player; //nothing changes Leave as Ace
        break;
      case 1:
        player = player - 10;
        break;
      }
      if ((player > dealer) && (dealerSticks == 1)){
        win = 1;                        //dealer sticks and you win
        win_lose = 2;
      }
```

```
    //displays values after Ace selection
    Serial.println (" ");
    Serial.print ("OK Your Total is ");
    Serial.print (player);
    Serial.print (", and the Dealer's Total is ");
    Serial.println (dealer);
    delay (1000);
    ace = false;
    aceLatch = 0;
    Serial.println (" ");
  } // end of ace

  ///////////////////////////////////////  code for hits
  while ((player < 21) && (playerSticks == 0) && (win == 0) &&
(lose == 0)) {
    Serial.println (" ");
    delay (500);
    Serial.println ("Would You Like a Hit ?");
    Serial.println (" ");
    Serial.println ("(pin 2 if yes, pin 3 if no)");
    Serial.println (" ");
    timed = 0;
    while (timed < 6000){//loops while player decides hit or stay
      delay (10);
      if (timed == 3000){   //nag message
        Serial.println (" ");
        Serial.print ("If no decision in 30 seconds, hit = NO");
        Serial.println (" ");
      }
      if (timed == 4500){
        Serial.println (" ");
        Serial.println ("***Hit ? ***");
        Serial.println (" ");
      }
      if (timed == 5999){
        playerSticks = 1;
      }
      timed++;
    }
    delay (100);
    if ((playerSticks == 0) && (hitCounter < 3)){
```

```
  hitCounter++;
  Serial.println (" ");
  Serial.print ("Your next Card is ");//your third card, begin
  switch (randomNum5){
case 11:
  Serial.print ("an Ace, worth ");
  ace = 1;
  break;

case 12:
  Serial.print ("a Jack, worth ");
  break;
case 13:
  Serial.print ("a Queen, worth ");
  break;
case 14:
  Serial.print ("a King, worth ");
  break;
}
if (randomNum5 > 11){
  randomNum5 = 10;
}
Serial.print (randomNum5);
Serial.println (" ");
Serial.println (" ");
delay (1000);
player = player + randomNum5;
Serial.print ("Your Total is ");
Serial.print (player);
delay (1500);
Serial.println (" ");
Serial.println (" ");
delay (1000);
Serial.print ("The Dealer's Total is ");
Serial.println (dealer);
delay (1500);
Serial.println (" ");
if (ace == 1){           //player selects Ace as 1 or 10 points
  Serial.println (" ");
  Serial.print("??? Want the Ace to be 1 or 11 points ???");
```

```
      Serial.println (" *** (pin 2) is 11 points, (pin 3) is one
point ***");
      timed = 0;
      while (timed < 6000){
        delay (10);
        if (timed == 3000){     //nag
          Serial.println (" ");
          Serial.println ("Ace 11 or 1 point ?");
          Serial.println (" ");
          Serial.println ("If no decision it remains worth 11
points");
          Serial.println (" ");
        }
        if (timed == 4500){
          Serial.println (" ");
          Serial.println ("Ace 11 or 1 point ?");
          Serial.println (" ");
        }
        if (timed == 5999){
          aceLatch = 10;
        }
        timed++;
      }
      switch (aceLatch){
      case 10:
        player = player; //nothing changes Leave as Ace
        break;
      case 1:
        player = player - 10;
        break;
      }
    } //end of ace
  }     // end of player sticks
  if (aceLatch == 1){
    Serial.println (" ");
    Serial.println (" ");
    Serial.print ("OK Your Total is ");
    Serial.println (player);
  }
  delay (1500);
  ace = 0;
  aceLatch = 0;
  if ((hitCounter == 3) && (player < 22)){ //5 cards
```

```
     win = 1;
     hits = 1;
   }   //end of 5 card win
 } // end of player under 21
  if (player > 21){        /////player over 21
  win_lose = 4;
  lose = 1;
  }
  while ((dealerSticks == 0) && (dealer < player) && (dealer < 21)
&& (win == 0) && (lose == 0)){
                              //Dealer's next card
    randomNum6 = random (2, 15);
    Serial.println (" ");
    Serial.print("The Dealer's Next Card is ");//dealer's next card
    switch (randomNum6){
    case 11:
      Serial.print ("an Ace worth ");
      ace = 1;
      break;
    case 12:
      Serial.print ("a Jack worth ");
      break;
    case 13:
      Serial.print ("a Queen worth ");
      break;
    case 14:
      Serial.print ("a King worth ");
      break;
    }
    delay (10);
    if (randomNum6 > 11){
      randomNum6 = 10;
    }
    dealer = dealer + randomNum6;
    if ((dealer > 21) && (ace == 1)){
      randomNum6 = 1;
      dealer = dealer - 10;
      ace = 0;
      aceLatch = 0;
    }
    Serial.print (randomNum6);
    Serial.println (" ");
```

```
  delay (2000);
  Serial.print ("Dealer total = ");
  Serial.print (dealer);
  Serial.println (" ");  // dealer's next card, end code:
  ace = 0;
  aceLatch = 0;
  delay (1500);
}
if ( hits == 1){  //5 card win
  win_lose = 3;
  win = 1;
}
if ((player > dealer) && (player < 22) && (hits != 1)){//player
                                                    //wins
  win_lose = 5;
  win = 1;
}
if ((dealer >= player) && (dealer < 22)){ //dealer beats player
  win_lose = 6;
  lose = 1;
}
if (dealer > 21){         //dealer busted
  win_lose = 7;
  win = 1;
}   //****************
switch (win_lose){
case 1:
  Serial.println (" ");
  Serial.println (" ");
  Serial.print ("The Dealer Gets BlackJack, You Lose");
  delay (500);
  Serial.println (" ");
  Serial.println (" ");
  break;
case 2:
  Serial.println (" ");//dealer sticks,player higher without hit
  Serial.println (" ");
  Serial.print ("Dealer sticks, and Player Wins");
  delay (500);
  Serial.println (" ");
```

```
  Serial.println (" ");
  break;
case 3:
  Serial.println (" ");      //5 card win
  Serial.println (" ");
  Serial.print ("Player Wins with Five Cards");
  delay (500);
  Serial.println (" ");
  Serial.println (" ");
  break;
case 4:
  Serial.println (" ");  //player goes over 21
  Serial.println (" ");
  Serial.print ("PLAYER BUSTED");
  delay (500);
  Serial.println (" ");
  Serial.println (" Sorry, You Lose ");
  Serial.println (" ");
  break;
case 5:
  Serial.println (" ");      //player gets a higher score
  Serial.println (" ");
  Serial.print ("! Player beats Dealer !");
  delay (500);
  Serial.println (" ");
  Serial.println (" ");
  break;
case 6:
  Serial.println (" ");      //dealer beats player
  Serial.println (" ");
  Serial.print ("! Dealer beats Player !");
  delay (500);
  Serial.println (" ");
  Serial.println (" ");
  break;
case 7:
  Serial.println (" ");          //dealer goes over 21
  Serial.println (" ");
  Serial.print ("! D E A L E R   B U S T E D !");
  delay (500);
  Serial.println (" ");
  Serial.println ("!!! You Win!!! ");
  Serial.println (" ");
```

```
      delay (500);
      break;
    }
    delay (10);
    trigState = 0;
    trigLatch = 0;
  } //end of trigLatch
} //end of main loop
//******** hardware interrupts *********

//must guard against switch bounce
void yes_ISR(){
  if (ace == true){
    aceLatch = 10;
    timed = 6000;
  }
  else{
    randomNum5 = random (2, 15);
    timed = 6000;
  }  // ace yes, and your next card end code:
}
void no_ISR(){
  if (ace == true){
    aceLatch = 1;
    timed = 6000;
  }
  else{
    playerSticks = 1;
    timed = 6000;
  }
}
```

Using the Arduino to Transmit Morse Code

Morse code is somewhat similar to the binary system, in that it also uses
two values, but rather than being logic levels, Morse signals are time
dependent. A *dot* is represented by a signal of a short duration, whereas a
dash is represented by a signal of a longer duration. Each Morse signal is

separated by one dot time period. The differentiation of the letters within words are also time dependent, as are the groupings of complete words. The dot is the basic unit of time. The other binary case is the dash, which consists of a signal of three dot time durations. The 26 individual letters of the English alphabet consist of different groupings of dots and dashes, as do the numbers 0 through 9. There is no provision for uppercase or lowercase letters. Originally Morse code was used for wired telegraph communication and migrated to radio when wireless communication was developed in the late 19th century. Soon thereafter, it was discovered how normal audio could be transmitted over radio. The first commercial radio license was issued to KDKA in Pittsburgh, Pennsylvania, in 1920. It featured music, news, and entertainment programs. Before that time, Morse code was used regularly as the communication method for ships and even for point-to-point communication over land. The use of Morse code still has an advantage of power efficiency, because even under poor broadcast conditions, it is easy to hear the individual dots and dashes. Although it is no longer a license requirement for amateur radio operators, some still use it regularly as it has the benefit of cutting through harsh reception conditions when voice signals could be misunderstood or unintelligible.

The code in our final regular project (Listing 9-3) looks daunting, but much of the code is repetitive and can be easily copied and pasted. As with many of the code examples in this book, it is available as a download. The program allows a user to type in sentences with up to 64 characters. The UNO maximum buffer size is 64 bytes and anything more is ignored and lost. At the end of each sentence when the user presses the Enter key or clicks the Send button on the serial monitor IDE screen, the Morse code signals are sent letter by letter, and the text is displayed on the IDE screen as it is transmitting. The default speed can be altered by changing the delay times in the declaration section at the top of the code. The time periods we are using will produce a transmission speed of about five words per minute, which is a good beginning speed. To monitor the output of the Arduino, an LED can be connected to pin 7 through a current-limiting resistor of around 220 Ohms, but it would be better to use a speaker connected to pin 7 (remember to use a series resistor larger than 100 Ohms). To actually transmit the data over a radio, an interface would need to be constructed.

Table 9-1 shows the Morse code dot and dash patterns for the representation of letters and numbers. It helps to represent the dots and dashes graphically. An upcoming capstone project is to write a program to receive data in Morse form and convert it to text. Many programs already exist in the amateur radio community to do both the transmission and reception of Morse code, but it would be interesting to do it from scratch and to customize the programming code for specific applications.

Table 9-1. *Morse Code*

a =	● ▬		s =	● ● ●
b =	▬ ● ● ●		t =	▬
c =	▬ ● ▬ ●		u =	● ● ▬
d =	▬ ● ●		v =	● ● ● ▬
e =	●		w =	● ▬ ▬
f =	● ● ▬ ●		x =	▬ ● ● ▬
g =	▬ ▬ ●		y =	▬ ● ▬ ▬
h =	● ● ● ●		z =	▬ ▬ ● ●
I =	● ●		1 =	● ▬ ▬ ▬ ▬
J =	● ▬ ▬ ▬		2 =	● ● ▬ ▬ ▬
k =	▬ ● ▬		3 =	● ● ● ▬ ▬
l =	● ▬ ● ●		4 =	● ● ● ● ▬
m =	▬ ▬		5 =	● ● ● ● ●
n =	▬ ●		6 =	▬ ● ● ● ●
o =	▬ ▬ ▬		7 =	▬ ▬ ● ● ●
p =	● ▬ ▬ ●		8 =	▬ ▬ ▬ ● ●
q =	▬ ▬ ● ▬		9 =	▬ ▬ ▬ ▬ ●
r =	● ▬ ●		0 =	▬ ▬ ▬ ▬ ▬

You can breadboard three momentary switches or use the process of quickly tapping a wire to ground to change the code transmission speed. The default speed can be changed by tapping pin 9 to ground so that the code speed steps up to approximately 10 words per minute, and pin 10 provides about 15 words per minute. At any time, you can reset the speed back to five words per minute by momentarily grounding pin number 8. Notice that we are using the dot time as our reference and adjusting the other characteristics accordingly.

Listing 9-3. Program for Transmitting Mose Code

```
int keys;//generates Morse code from sentences typed in serial monitor
int line; //also displays letters on screen as they are sent
int ASCII;//the delay function starts at the beginning of the tone
int dot = 250;//we use 250 ms as dot time
int dash; //a dash is 3 dot times.
int space; //space between letters 3 dot times
int wordSpace;//7 dot times between words
boolean fast;
boolean med;
boolean slow;
void setup () {
pinMode (10, INPUT_PULLUP);
pinMode (9, INPUT_PULLUP);
pinMode (8, INPUT_PULLUP);
pinMode (7, OUTPUT);
Serial.begin (9600);
Serial.println ("This program will transmit Morse code");
Serial.println ("up to 64 characters per line (about 10 words)");
Serial.println ("default speed is 5 WPM");
Serial.println ("pin 9 gives 10 WPM, and pin 10 is for 15 WPM");
Serial.println ("tapping pin 8 to ground returns to 5 WPM.");
Serial.println ("***Type text and enter***");
}
```

```
void loop () {
  fast = digitalRead (10);
  med = digitalRead (9);
  slow = digitalRead (8);
  if (fast == LOW) {
    dot = 83;
  }
  else if (med == LOW) {
    dot = 125;
  }
  else if (slow == LOW) {
    dot = 250;
  }
  dash = 3 * dot;
  space = dash;
  wordSpace = (7 * dot) - space; //subtracting the letter delay

  while (Serial.available () > 0) {//grabs a byte from the buffer
    line = 1;
    keys = Serial.read ();

    switch (keys) {       //ASCII decoding section
    case 32:       //space bar case, detects space between words
      delay (wordSpace); //seven dot time periods between words
      break;
//*********** Numbers ************
case 48:     //number 0
      tone (7, 2000, dash);
      delay(dash + dot);
      tone (7, 2000, dash);
      delay (dash + dot);
      tone (7, 2000, dash);
      delay(dash + dot);
      tone (7, 2000, dash);
      delay (dash + dot);
      tone (7, 2000, dash);
      delay (dash + space);
      break;

  case 49:     //number 1
      tone (7, 2000, dot);
      delay(dot + dot);
      tone (7, 2000, dash);
```

```
  delay(dash + dot);
  tone (7, 2000, dash);
  delay (dash + dot);
  tone (7, 2000, dash);
  delay(dash + dot);
  tone (7, 2000, dash);
  delay (dash + space);
  break;
case 50:     //number 2
  tone (7, 2000, dot);
  delay (dot + dot);
  tone (7, 2000, dot);
  delay (dot + dot);
  tone (7, 2000, dash);
  delay (dash + dot);
  tone (7, 2000, dash);
  delay(dash + dot);
  tone (7, 2000, dash);
  delay (dash + space);
  break;
case 51:     //number 3
  tone (7, 2000, dot);
  delay (dot + dot);
  tone (7, 2000, dot);
  delay (dot + dot);
  tone (7, 2000, dot);
  delay (dot + dot);
  tone (7, 2000, dash);
  delay(dash + dot);
  tone (7, 2000, dash);
  delay (dash + space);
  break;
case 52:     //number 4
  tone (7, 2000, dot);
  delay (dot + dot);
  tone (7, 2000, dot);
  delay (dot + dot);
  tone (7, 2000, dot);
  delay (dot + dot);
  tone (7, 2000, dot);
  delay (dot + dot);
  tone (7, 2000, dash);
  delay(dash + space);
  break;
```

```
case 53:      //number 5
  tone (7, 2000, dot);
  delay (dot + dot);
  tone (7, 2000, dot);
  delay (dot + dot);
  tone (7, 2000, dot);
  delay (dot + dot);
  tone (7, 2000, dot);
  delay (dot + dot);
  tone (7, 2000, dot);
  delay (dot + space);
  break;
case 54:      //number 6
  tone (7, 2000, dash);
  delay (dash + dot);
  tone (7, 2000, dot);
  delay (dot + dot);
  tone (7, 2000, dot);
  delay (dot + dot);
  tone (7, 2000, dot);
  delay (dot + dot);
  tone (7, 2000, dot);
  delay (dot + space);
  break;
case 55:      //number 7
  tone (7, 2000, dash);
  delay(dash + dot);
  tone (7, 2000, dash);
  delay (dash + dot);
  tone (7, 2000, dot);
  delay (dot + dot);
  tone (7, 2000, dot);
  delay (dot + dot);
  tone (7, 2000, dot);
  delay (dot + space);
  break;
case 56:      //number 8
  tone (7, 2000, dash);
  delay(dash + dot);
  tone (7, 2000, dash);
  delay (dash + dot);
  tone (7, 2000, dash);
  delay(dash + dot);
```

```
   tone (7, 2000, dot);
   delay (dot + dot);
   tone (7, 2000, dot);
   delay(dot + space);
   break;
 case 57:    //number 9
   tone (7, 2000, dash);
   delay(dash + dot);
   tone (7, 2000, dash);
   delay (dash + dot);
   tone (7, 2000, dash);
   delay(dash + dot);
   tone (7, 2000, dash);
   delay (dash + dot);
   tone (7, 2000, dot);
   delay (dot + space);
   break;
// ********** Letters **********
 case 97:  // letter a  -  lowercase
 case 65: // letter a  -  uppercase
   tone (7, 2000, dot);
   delay(dot + dot);
   tone (7, 2000, dash);
   delay(dash + space);
   break;
 case 98:            // letter b
 case 66:
   tone (7, 2000, dash);
   delay (dash + dot);
   tone (7, 2000, dot);
   delay (dot + dot);
   tone (7, 2000, dot);
   delay (dot + dot);
   tone (7, 2000, dot);
   delay(dot + space);
   break;
 case 99:         //letter c
 case 67:
   tone (7, 2000, dash);
   delay (dash + dot);
   tone (7, 2000, dot);
   delay (dot + dot);
   tone (7, 2000, dash);
   delay (dash + dot);
```

```
  tone (7, 2000, dot);
  delay (dot + space);
  break;
case 100:          //letter d
case 68:
  tone (7, 2000, dash);
  delay (dash + dot);
  tone (7, 2000, dot);
  delay (dot + dot);
  tone (7, 2000, dot);
  delay (dot + space);
  break;
case 101:          //letter e
case 69:
  tone (7, 2000, dot);
  delay (dot + space);
  break;
case 102:          //letter f
case 70:
  tone (7, 2000, dot);
  delay (dot + dot);
  tone (7, 2000, dot);
  delay (dot + dot);
  tone (7, 2000, dash);
  delay (dash + dot);
  tone (7, 2000, dot);
  delay (dot + space);
  break;
case 103:          //letter g
case 71:
  tone (7, 2000, dash);
  delay (dash + dot);
  tone (7, 2000, dash);
  delay (dash + dot);
  tone (7, 2000, dot);
  delay (dot + space);
  break;
case 104:          //letter h
case 72:
  tone (7, 2000, dot);
  delay (dot + dot);
  tone (7, 2000, dot);
  delay (dot + dot);
```

```
  tone (7, 2000, dot);
  delay (dot + dot);
  tone (7, 2000, dot);
  delay (dot + space);
  break;
case 105:           //letter i
case 73:
  tone (7, 2000, dot);
  delay (dot + dot);
  tone (7, 2000, dot);
  delay (dot + space);
  break;
case 106:           //letter j
case 74:
  tone (7, 2000, dot);
  delay(dot + dot);
  tone (7, 2000, dash);
  delay(dash + dot);
  tone (7, 2000, dash);
  delay (dash + dot);
  tone (7, 2000, dash);
  delay (dash + space);
  break;
case 107:         //letter k
case 75:
  tone (7, 2000, dash);
  delay (dash + dot);
  tone (7, 2000, dot);
  delay (dot + dot);
  tone (7, 2000, dash);
  delay (dash + space);
  break;
case 108:         //letter l
case 76:
  tone (7, 2000, dot);
  delay(dot + dot);
  tone (7, 2000, dash);
  delay(dash + dot);
  tone (7, 2000, dot);
  delay (dot + dot);
  tone (7, 2000, dot);
  delay (dot + space);
  break;
```

```
case 109:          //letter m
case 77:
  tone (7, 2000, dash);
  delay (dash + dot);
  tone (7, 2000, dash);
  delay (dash + space);
  break;
case 110:          //letter n
case 78:
  tone (7, 2000, dash);
  delay (dash + dot);
  tone (7, 2000, dot);
  delay (dot + space);
  break;
case 111:          //letter o
case 79:
  tone (7, 2000, dash);
  delay(dash + dot);
  tone (7, 2000, dash);
  delay (dash + dot);
  tone (7, 2000, dash);
  delay (dash + space);
  break;
case 112:          //letter p
case 80:
  tone (7, 2000, dot);
  delay(dot + dot);
  tone (7, 2000, dash);
  delay(dash + dot);
  tone (7, 2000, dash);
  delay (dash + dot);
  tone (7, 2000, dot);
  delay (dot + space);
  break;
case 113:          //letter q
case 81:
  tone (7, 2000, dash);
  delay(dash + dot);
  tone (7, 2000, dash);
  delay (dash + dot);
  tone (7, 2000, dot);
  delay(dot + dot);
  tone (7, 2000, dash);
```

```
  delay(dash + space);
  break;
case 114:           //letter r
case 82:
  tone (7, 2000, dot);
  delay(dot + dot);
  tone (7, 2000, dash);
  delay(dash + dot);
  tone (7, 2000, dot);
  delay (dot + space);
  break;
case 115:           //letter s
case 83:
  tone (7, 2000, dot);
  delay (dot + dot);
  tone (7, 2000, dot);
  delay (dot + dot);
  tone (7, 2000, dot);
  delay (dot + space);
  break;
case 116:           //letter t
case 84:
  tone (7, 2000, dash);
  delay(dash + space);
  break;
case 117:           //letter u
case 85:
  tone (7, 2000, dot);
  delay (dot + dot);
  tone (7, 2000, dot);
  delay (dot + dot);
  tone (7, 2000, dash);
  delay (dash + space);
  break;
case 118:           //letter v
case 86:
  tone (7, 2000, dot);
  delay (dot + dot);
  tone (7, 2000, dot);
  delay (dot + dot);
  tone (7, 2000, dot);
  delay (dot + dot);
  tone (7, 2000, dash);
```

```
      delay(dash + space);
      break;
   case 119:          //letter w
   case 87:
      tone (7, 2000, dot);
      delay(dot + dot);
      tone (7, 2000, dash);
      delay(dash + dot);
      tone (7, 2000, dash);
      delay(dash + space);
      break;
   case 120:          //letter x
   case 88:
      tone (7, 2000, dash);
      delay (dash + dot);
      tone (7, 2000, dot);
      delay (dot + dot);
      tone (7, 2000, dot);
      delay (dot + dot);
      tone (7, 2000, dash);
      delay(dash + space);
      break;
   case 121:          //letter y
   case 89:
      tone (7, 2000, dash);
      delay (dash + dot);
      tone (7, 2000, dot);
      delay (dot + dot);
      tone (7, 2000, dash);
      delay (dash + dot);
      tone (7, 2000, dash);
      delay(dash + space);
      break;
   case 122:          //letter z
   case 90:
tone (7, 2000, dash);
      delay(dash + dot);
      tone (7, 2000, dash);
      delay (dash + dot);
      tone (7, 2000, dot);
      delay(dot + dot);
      tone (7, 2000, dot);
      delay (dot + space);
```

```
    break;
  }
  Serial.print (keys, ASCII);
}//now all characters in buffer have been sent and displayed
if (line == 1){
  Serial.println (); //prints next sentence on a new line
  line = 0;
}
}
```

There are many different ways to code a program. It is an exercise in problem solving. First, we must understand the problem. The second step is to find an overarching strategy to tackle the problem. Finally, tactics are devised to find a correct solution to the problem. In coding, we have the additional headache of entering the correct programming language syntax into the computer. It is highly advisable to take time and put much effort into the problem-solving aspect of finding a solution, as finding syntax errors is somewhat more straightforward. Spending time before doing any actual coding by drawing sketches and diagrams, drawing flowcharts, or writing pseudocode out by hand will be well worth the effort. Chapter 10 outlines projects that will ask you to devise solutions to make modifications to given code. It might be helpful to revisit some of the programs presented in earlier sections of the text. The reference section in the Arduino IDE is also a very handy source of information.

CHAPTER 10

Capstone Projects

Building an Audio Morse Code Reader

Throughout the text, we have presented projects that demonstrated a wide variety of coding techniques and asked you to make modifications to gain a greater understanding of the programming process. In this capstone chapter, we present projects that are fairly difficult and require you to improvise solutions to produce a fully functioning project. This first capstone section asks you to build a reader that will respond to the last project presented in Chapter 9. The electronic circuit and corresponding Arduino code will read the Morse code to some extent. The code as given will respond to the slow (default) code speed, and to the audio frequency of 2 kHz sent to a small speaker. When the speaker is held within a few inches of the microphone, the code that is sounded is converted into information on the reader program's serial monitor. The program only converts the sounds back to the following symbols: dot, dash, letter, word, and a hyphen that is shown between the dots and dashes. To recover this information, the code responds to the sound and finds the total time of each tone by adding short increments of sequential samples. It does this by using the `goto` function, which rotates around a section of code while the audio tone is present. It redirects the program operation to the start of a section of code that we call `sampleStart:`. It should be noted that `goto` commands are frowned on by experienced programmers because they tend to make programs difficult to follow. It works nicely in our program, but could be replaced with the `do` or `do-while` commands. This would

© Bob Dukish 2018
B. Dukish, *Coding the Arduino*, https://doi.org/10.1007/978-1-4842-3510-2_10

necessitate having do functions within do functions, however, which might be considered as nested do-do, and could make the code even messier.

In the code that is provided, we also use a subroutine (sometimes called a function) to find and print out the transmitted information. (Subroutine use in C and C++ is discouraged, but sometimes can help eliminate the need to repeat sections of code.) In our code, we move to the subroutine called routine, which is located outside of the main loop, to check for spacing times.

By using the slow code speed, the information is decoded by the following time periods: 250 ms represents a dot and three dot times represent a dash. For Morse code spacing, the following convention is used: The space between dots and dashes is the same as for a letter; at our code speed, this is one dot time (250 ms). The spaces between complete letters and numbers is three dot times and the space between words is seven dot times. The circuit that we are using is simple, inexpensive, and functional. If any difficulties are encountered, it is advisable to use a small 8 Ohm speaker of 2- to 3-inch diameter and hold it steady within a distance from 3 to 5 inches from the microphone. Tests should be run in a quiet environment. If there is no response or an incorrect response, the code transmitter might need additional amplification. Once the given code is tested with your circuit, the objective of this capstone project is to modify the given code so that it can respond to the other two transmitting speeds from the project shown in Chapter 9.

The microphone circuit (Figure 10-1) uses an LM386 amplifier set with the maximum gain of 200. The 386 is a power amplifier best used to drive small speakers, but because of its low cost and availability it is used instead of an op-amp in this project. The 386 output is coupled to an analog input on the Arduino, where the 1 k Ohm voltage divider sets the DC level exactly at 2.5 volts DC with no audio. (No audio from a 386 has a DC level of 2.5 volts, but sometimes it can be a little off.) The DC offset is used to bring the audio signal above ground and avoid clipping the negative portion of the sine wave. The Arduino interprets voltages at its analog input pins at values of 0 up to 1,023, which corresponds to DC voltages of from 0 to 5 volts. Therefore, each volt has an Arduino analog representative value of approximately 205. With no audio input, our voltage divider sets the number to approximately 512. As the audio from the tone causes the wave to rise above and below the reference, which represents zero for no sound, the half-waves are counted in a manner similar to the frequency counter project in Chapter 8. The program code looks for audio time periods to identify the dots and dashes, and for silent time periods to find the spacing between them.

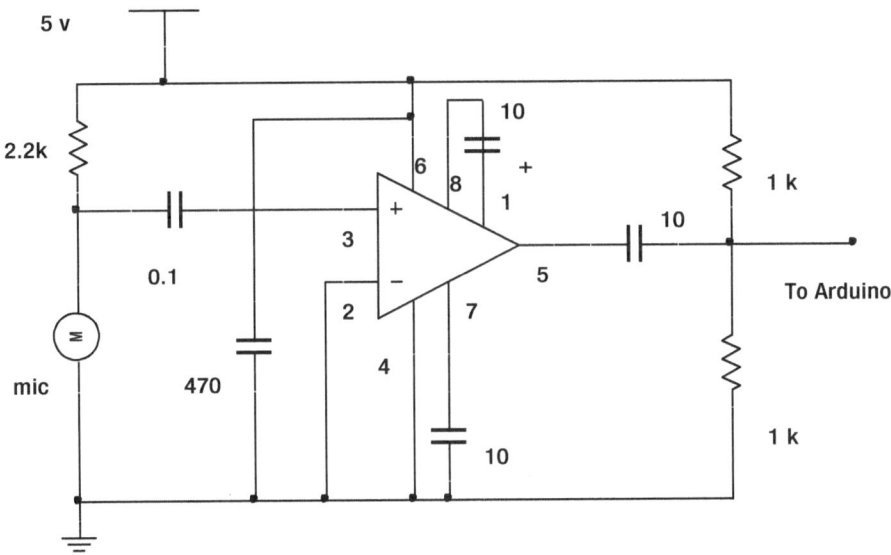

Figure 10-1. *Schematic of the mic amp circuit with a gain of 200 using an LM386*

It is recommended that two Arduino boards be used for this project; however, one board could be used if it is first configured as an audio transmitter and a line of Morse code is then recorded on another device such as a smartphone. The board can subsequently be set up as the receiver, so that it can then respond to the prerecorded sounds. It is best to proceed in small sections at a time, such as getting the circuit and program working with a few recognizable characters such as repeating the phrase SOS, which has the code "dit-dit-dit (letter) dash-dash-dash (letter) dit-dit-dit" (word). It is folklore that the letters SOS stand for "save our ship." The actual reason it was chosen as the Morse code distress call was due to its recognizable pattern.

This is a challenging capstone project. Commercial programs have been written for ham radio operators that will go to the next step, which is to decode the dots, dashes, and spacing back into the original text. Our next project will make our program more complete. The main objective of this project (Listing 10-1) is for you to modify the given code so that it will be switchable to allow operation at the three different speeds that our transmitter can send.

Listing 10-1. An Audio Morse Code Reader

```
                    //code reader program
unsigned long currentMillis; //First establishes the tone frequency
unsigned long lastMillis;    //then finds the dot time
unsigned int duration;       //and uses Morse transmission spacing to
int pulseHigh;               //recover coded letters and words
int pulseLow;
long halfCycles;
long cycles;
int in;
unsigned int pulseTime;
unsigned long startNext;
unsigned long stopLast;
int space;
```

```
unsigned long wordEnd;
unsigned int wordLast;
boolean dotDash;

void setup(){
  Serial.begin(9600);
  Serial.println("Welcome to the Code Reader Program");
}

void loop(){
sampleStart:
  cycles = 0;
  halfCycles = 0;
  lastMillis = millis();

do{
  in = analogRead(A0);//connect output of circuit to the analog pin
  currentMillis = millis();
  duration = currentMillis - lastMillis; //loops for 10 ms
    if (in > 540){
    pulseHigh = 1;
  }
  if (in < 500){
    pulseLow = 1;
  }
  if (pulseHigh == 1 && pulseLow == 1){
    halfCycles = halfCycles + 1;
    pulseHigh = 0;
    pulseLow = 0;
  }
}
while(duration < 10);     //counts half cycles for 10 ms
                 //multiply by 100 for cycles per second (cps)
cycles = (100)*(halfCycles / 2);//divide by two there are two half
                                //cycles, per cycle
if (cycles > 500 && cycles < 3500){ //looks for an approx 2 khz tone
  pulseTime = pulseTime + 1; //each pulseTime period is 10 ms
  goto sampleStart;       //go back to sample while tone present
}
```

```
    pulseTime = pulseTime * 10;//times ten for 10 ms of elapsed time
    if (pulseTime > 150 && pulseTime < 350) { //looks for dots
      stopLast = millis();
      space = stopLast - startNext;
      space = space - 250; //subtract dot time
      routine();
      startNext = millis();
      pulseTime= 0;
      dotDash = 1;
      Serial.print ("dot");
    }
    if (pulseTime > 650 && pulseTime < 850){ //looks for dashes
      stopLast = millis();
      space = stopLast - startNext;
      space = space - 750; //subtract dash time
      routine();
      startNext = millis();
      pulseTime= 0;
      dotDash = 1;
      Serial.print ("dash");
    }
    wordEnd = millis();      //identifies the last word
    wordLast = wordEnd - startNext;
    if ((wordLast > 1850 && wordLast < 1900) && (dotDash == 1)){
      Serial.print (" word ");
      dotDash = 0;
    }
}
/////////////////////////subroutine////////////////////
void routine(){
  if (space > 200 && space < 300){
    Serial.print ("-");
  }
  if (space > 700 && space < 800){
    Serial.print (" letter ");
  }
  if (space > 1700 && space < 1800 ){
    Serial.print (" word ");
  }
}
```

Building an Audio Morse Code Decoder

In this project (Listing 10-2), we adapt the program code from the previous section to decode the transmitted audio from the Morse code audio transmitter presented in Chapter 8. It operates in a very similar fashion to the first capstone project, and you will need to incorporate the ability to work with differing transmission speeds, which was the requirement of the last project. Additionally, you will need to develop a database of Morse characters that will be compared with the received information, to generate letters and numbers on the program's serial monitor, which will match the transmitted text.

The operation of the given program code should be tested first. Much of it can be copied and pasted from the last program into the new code that you are developing. The Morse strings for the letters S and O are given, and the others are listed in Table 9-1 in Chapter 9. As in the last project, you will need two Arduino boards, or you can use just one, by first running the audio transmit program and using a recording device such as a smartphone, and later playing back the sounds as this project's receiver program is running. The objective of the capstone project is for you to complete the code, so that as the tones are generated by the transmitter and subsequently received by the microphone circuit (Figure 10-1), the program will display the information as text on the serial monitor.

Listing 10-2. An Audio Morse Code Decoder

```
            //code decoder program for Morse code
unsigned long currentMillis; //First establishes the tone frequency
unsigned long lastMillis;    //then finds the dot time
unsigned int duration;    //and uses Morse transmission spacing to
int pulseHigh;            //recover coded letters and words
int pulseLow;
long halfCycles;
```

```
long cycles;
int in;
unsigned int pulseTime;
unsigned long startNext;
unsigned long stopLast;
int space;
unsigned long wordEnd;
unsigned int wordLast;
boolean dotDash;
String strOne;
int wordNumber;
void setup(){
  Serial.begin(9600);
  Serial.println("Welcome to the Code Reader Program");
}
void loop(){
sampleStart:
  cycles = 0;
  halfCycles = 0;
  lastMillis = millis();
  do{
    in = analogRead(A0); //output of the circuit to the analog pin
    currentMillis = millis();
    duration = currentMillis - lastMillis; //loops for 10 ms
      if (in > 540){
      pulseHigh = 1;
    }
    if (in < 500){
      pulseLow = 1;
    }
    if (pulseHigh == 1 && pulseLow == 1){
      halfCycles = halfCycles + 1;
      pulseHigh = 0;
      pulseLow = 0;
    }
  }
  while(duration < 10);     //counts half cycles for 10 ms
                  //multiply by 100 for cycles per second (cps)
```

```
cycles = (100) * (halfCycles / 2);//two half per cycle
if (cycles > 500 && cycles < 3500){ //looks for an approx 2 khz tone
  pulseTime = pulseTime + 1; //each pulseTime period is 10 ms
  goto sampleStart; // goes back, to sample while tone is present
}
pulseTime = pulseTime * 10; //times ten for 10 ms of elapsed time
if (pulseTime > 150 && pulseTime < 350) { //looks for dots
  stopLast = millis();
  space = stopLast - startNext;
  space = space - 250; //subtract dot time
  routine();
  startNext = millis();
  pulseTime= 0;
  dotDash = 1;
  strOne = strOne + "dot";
}
if (pulseTime > 650 && pulseTime < 850){ //looks for dashes
  stopLast = millis();
  space = stopLast - startNext;
  space = space - 750; //subtract dash time
  routine();
  startNext = millis();
  pulseTime= 0;
  dotDash = 1;
  strOne = strOne + "dash";
}
wordEnd = millis();     //identifies the last word
wordLast = wordEnd - startNext;
if ((wordLast > 1850 && wordLast < 1950) && (dotDash == 1)){
  routineLetters();
  strOne = ("");
  Serial.println (""); //"ln" makes a new line after last word
  dotDash = 0;
}
```

```
  if (wordNumber > 32) {
    Serial.println ("");//"ln" makes a new line to keep on pc screen
    wordNumber = 0;
  }
}
/////////////////////subroutines////////////////////
void routine(){
  if (space > 700 && space < 800){ //letter space
    routineLetters();
    strOne = ("");
  }
  if (space > 1700 && space < 1800 ){ //word space
    routineLetters();
    strOne = ("");
    Serial.print (" ");
    wordNumber = wordNumber + 1;
  }
}
void routineLetters(){
  if (strOne == "dotdotdot"){
    Serial.print ("S");
    strOne = ("");
  }
  else if (strOne == "dashdashdash"){
    Serial.print ("O");
    strOne = ("");
  }
}
```

Team Project 1: IR Morse Code Link

This capstone project allows for text messages to be sent over an infrared (IR) link. We describe a one-way system, but the project could easily be developed into a two-way transceiver system. The project could also be adapted for communication over radio frequency links using nonlicensed 433 MHz transmitters and receivers, or over Wi-Fi using low-cost wireless transceivers such as the popular NRF24L01, which operates in the 2.4 GHz band. Our project uses the IR receiver module DFR0094, responsive to 38 KHz pulses. Any similar IR receiver module can be substituted, as long as it provides a logic low output condition while it is receiving 38 KHz pulses; otherwise the logic state needs to be a high. Any compatible IR LED can be used on the transmit side. For very close range (about 5 feet), the LED and a 120 Ohm resistor can be connected in series between output pin 7 and Arduino ground. For a range of about 20 feet, the circuit in Figure 10-2 can be used or modified with two or more LED circuits to increase the range even farther.

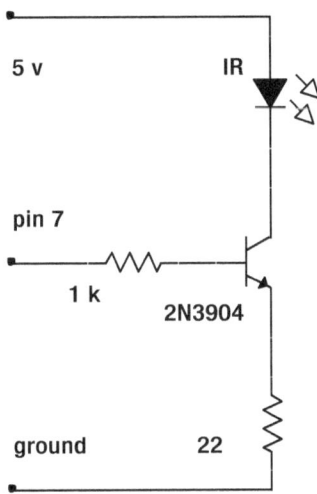

Figure 10-2. *A circuit used to increase range*

A 2N3904 NPN transistor is capable of a maximum collector current of 200 mA. By using a 22 Ohm emitter resistor in the circuit as shown, the continuous collector current would be 130 mA, and resistor power would be 0.4 watt, but we are operating at less than a 50% duty cycle so a quarter-watt resistor will not be overheated.

The transmitter code only needs to be slightly modified from what is shown in Chapter 9. The difference is that we replace the 2,000 Hz tone with a carrier pulse frequency of 38,000 Hz, which is the standard carrier frequency for IR devices. Because this value is larger than the maximum integer value, we assign the variable called led the unsigned int type in the top declaration section. Then at the beginning of the main loop, we add the code led = 38000; this section is shown in Listing 10-3, with the changes to the project highlighted.

Listing 10-3. Infrared Morse Code Link

```
int keys; //generates Morse code from sentences typed in the serial
monitor
int line;
int ASCII;
int dot = 250;
int dash;
int space;
int wordSpace;
boolean fast;
boolean med;
boolean slow;
unsigned int led;

void setup () {
pinMode (10, INPUT_PULLUP);
pinMode (9, INPUT_PULLUP);
pinMode (8, INPUT_PULLUP);
pinMode (7, OUTPUT);

Serial.begin (9600);
Serial.println ("This program will transmit Morse code");
Serial.println ("up to 64 characters per line (about 10 words)");
Serial.println ("default speed is 5 WPM");
Serial.println ("pin 9 gives 10 WPM, and pin 10 is for 15 WPM");
Serial.println ("tapping pin 8 to ground returns to 5 WPM.");
Serial.println ("***Type text and enter***");
}
void loop (){
  led = 38000;
  fast = digitalRead (10);
  med = digitalRead (9);
  slow = digitalRead (8);
```

Also, due to the internal construction of the timers in the ATmega328P processor used in the Arduino UNO, there are timing issues that result in using the tone function at high frequencies with the coding method that we previously used in Chapter 9. All of the ASCII cases for transmitting both numbers and letters need to be redone by calling the noTone function in our switch case section. As an example (Listing 10-4), the new revised code is shown for transmitting the S and O letters.

Listing 10-4. Revised Code for Transmitting S and O

```
case 111:          //letter O
case 79:
  tone (7, led);
  delay(dash);
  noTone(7);
  delay(dot);
  tone (7, led);
  delay (dash);
  noTone(7);
  delay(dot);
  tone (7, led);
  delay (dash);
  noTone(7);
  delay(space);
  break;
case 115:          //letter S
  case 83:
  tone (7,led);
  delay (dot);
  noTone(7);
  delay(dot);
  tone (7, led);
  delay (dot);
  noTone(7);
  delay(dot);
  tone (7, led);
  delay(dot);
  noTone(7);
  delay (space);
  break;
```

It is again recommended to use the letters SOS for testing, and to then roll out the code changes to all of the other letters and numbers once the testing is successful. The best course of action would be to put all of this ASCII to text data in the form of a library and include it in the program, but as a learning experience, for now it might be more helpful to actually see the entire code.

The decoding program must be changed significantly and is shown in its entirety, but with only the letters S and O completely described. All other possibilities for letters and numbers must be added in the subroutine that we call `routineLetters()`. As in the requirements for the previous capstone projects, you will need to modify the code to switch the decode speed to match the transmission speed.

In the code (Listing 10-5), finding the dot and dash time duration is straightforward. For the spacing, we use a two-step process. Figure 10-3 might aid in understanding the timing concepts used in the code.

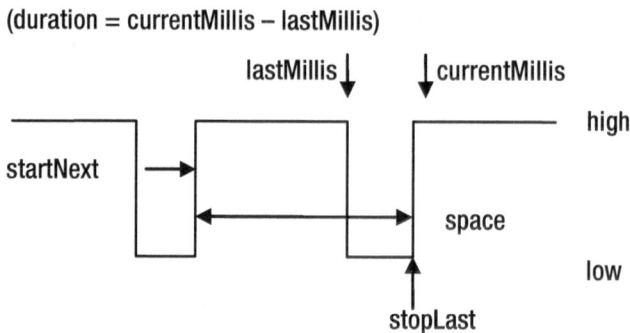

Figure 10-3. *Timing concepts used in the code*

Notice, in Listing 10-5, (`space = space - "the dot or dash duration time"`).

Listing 10-5. Infrared Morse Decoder

```
unsigned long currentMillis; //IR Morse code decode program
unsigned long lastMillis;
unsigned int duration;
boolean pulseHigh;
boolean pulseLow;
boolean in = HIGH;
unsigned long startNext;
unsigned long stopLast;
int space;
unsigned long wordEnd;
unsigned int wordLast;
boolean dotDash;
String strOne;
int wordNumber;
int dot;
int dash;
void setup(){
  Serial.begin(9600);
  Serial.println("Welcome to the IR Code Reader Program");
}
void loop(){
  dot = 250;
  dash = 3 * dot;
  in = digitalRead (7); // checks IR detector for a low
  if (in == LOW && pulseLow == 0){ //goes from high to low
    lastMillis = millis();
    pulseLow = 1;
  }
  if (in == HIGH && pulseLow == 1){ //goes from low to high
    currentMillis = millis();
    pulseHigh =1;
  }
  if (pulseLow == 1 && pulseHigh == 1){
    duration = currentMillis - lastMillis;
    pulseHigh = 0;
    pulseLow = 0;
  }
```

```
  if ((duration > (dot - dot/3)) && (duration < (dot + dot/3))){
  //looks for dots
    dotDash = 1;
    stopLast = millis();
    space = stopLast - startNext;
    space = space - dot; //subtract dot time
    routine();
    startNext = millis();
    strOne = strOne + "dot";
    duration = 0;
  }
  if (duration > (dash - dash/3) && duration < (dash + dash/3)){
//dashes
    dotDash = 1;
    stopLast = millis();
    space = stopLast - startNext;
    space = space - dash; //subtract dash time
    routine();
    startNext = millis();
    strOne = strOne + "dash";
    duration = 0;
  }
  wordEnd = millis();      //identifies the last word
  wordLast = wordEnd - startNext;
  if ((wordLast > (12*dot - dot/3) && wordLast < (12*dot + dot/3)) &&
    dotDash == 1){
    routineLetters();
    Serial.println (""); //"ln" makes a new line after last word
    dotDash = 0;
  }
  if (wordNumber > 32) {
    Serial.println ("");//"ln" makes a new line to keep on pc screen
    wordNumber = 0;
  }
}          ///////////////////////subroutines///////////////////////
void routine(){
  if (space > (3*dot - dot/3) && space < (3*dot + dot/3)){
//letter space
    routineLetters();//calls next subroutine, gets letter or number
  }
```

```
  if (space > (7*dot - dot/3) && space < (7*dot + dot/3)){ // word
    dotDash = 0;
    routineLetters();
    Serial.print (" "); //adds a space between words
    wordNumber = wordNumber + 1;
  }
}
void routineLetters(){
  if (strOne == "dotdotdot"){
    Serial.print ("S");
    strOne = ("");
  }
  else if (strOne == "dashdashdash"){
    Serial.print ("O");
    strOne = ("");
  }
}
```

Team Project 2: IR Control Link

Our final capstone project requiring electronic circuits uses two Arduino boards, with one programmed as a transmitter and the other as a receiver. We use IR pulses to convey information, but now we will be sending digital signals used to control devices on the receiver end of the link, the actual objective of what a microcontroller is meant to do. It is very similar to the way commercial IR remote controls work for most home entertainment products such as TVs and stereos. The commercial systems, however, will send a long series of pulses that make up a digital code for each of the device's functions.

The transmit IR LED schematic is shown in Figure 10-2, and the IR receiver can be the DFR0094 or any suitable substitute 38 kHz receiver module that gives a digital low output when the signal is received. The transmitter uses three input pins, which when momentarily grounded, will transmit a pulse train that corresponds to the code needed to activate a specific device on the receiver side of the link. We use three different color LEDs to simulate the devices. Because two Arduinos are needed, this makes a good team project. Two separate computers can be used, or just one can be used to program both boards provided that operational power is then supplied to the stand-alone board. This can be accomplished by connecting the V in the header pin to a voltage a few volts above 5 volts, so that the on-board regulator supplies the correct voltage to the Arduino. On the receiver end, the negative side (cathode) of the LEDs can each be connected though a 220 Ohm resistor to ground, or you can tie the three cathodes together and use just one 220 Ohm resistor connected to ground. The code in Listing 10-6 will send a 2 ms pulse at three different pulse repetition rates, depending on the selection pin that is put to ground. As a capstone project, you might wish to modify the code to switch the remote device both on and off. Additionally, you might wish to develop interfacing circuitry to control devices other than the LEDs used in this project.

Listing 10-6. Creating an IR Control Link

```
////////Code for the transmitter, Arduino board one
const int led1 = 8;
const int led2 = 9;
const int led3 = 10;
const int out = 7;
int A;
int B;
int C;
int light;
int timer;
void setup(){
  pinMode (led1, INPUT_PULLUP);
  pinMode (led2, INPUT_PULLUP);
  pinMode (led3, INPUT_PULLUP);
  pinMode (out, OUTPUT);
}
void loop(){
  A = digitalRead (led1);
  B = digitalRead (led2);
  C = digitalRead (led3);
```

```
  if (A == LOW){
    light = 1;
  }
  if (B == LOW){
    light = 2;
  }
  if (C == LOW){
    light = 3;
  }
  switch (light){
  case 1:
    while (timer < 10){
      tone (out, 38000);//outputs a short duration 38,000 tone
      delay (2);
      noTone (out);
      delay (8); //total period 10 ms
      timer = timer + 1;
    }
    break;
  case 2:
    while (timer < 10){
      tone (out, 38000);
      delay (2);
      noTone (out);
      delay (18); //period 20 ms
      timer = timer + 1;
    }
    break;
  case 3:
    while (timer < 10){
      tone (out, 38000);
      delay (2);
      noTone (out);
      delay (28); //period 30 ms
      timer = timer + 1;
    }
    break;
  }
  light = 0;
  timer = 0;
}
```

```
////////Code for the receiver, Arduino board two
const int in = 7;
const int led1 = 8;
const int led2 = 9;
const int led3 = 10;
int pulseTone;
int pulsePeriod;
unsigned long currentTime;
unsigned long oldTime;
int i;
void setup(){
  pinMode (in, INPUT_PULLUP);
  pinMode (led1, OUTPUT);
  pinMode (led2, OUTPUT);
  pinMode (led3, OUTPUT);
  Serial.begin(9600);//can use serial monitor to see pulse width
}
void loop(){
  pulseTone = digitalRead (in);
  if (pulseTone == LOW){
    currentTime = millis();
    pulsePeriod = currentTime - oldTime; //picks up the second burst
    oldTime = currentTime;
    if (pulsePeriod  > 6 && pulsePeriod  < 14){
      Serial.println (pulsePeriod);
      for (i; i < 100; i++){
        digitalWrite (led1, HIGH);
        delay (30); //turns on led for 3 seconds
      }
      digitalWrite (led1, LOW);
      pulsePeriod = 0;
    }
    if (pulsePeriod  > 16 && pulsePeriod  < 24){
      Serial.print (pulsePeriod);
      for (i; i < 100; i++){
        digitalWrite (led2, HIGH);
        delay (30); //turns on led for 3 seconds
      }
```

```
  digitalWrite (led2, LOW);
  pulsePeriod = 0;
}

if (pulsePeriod  > 26 && pulsePeriod  < 34){
  for (i; i < 100; i++){
    digitalWrite (led3, HIGH);
    delay (30); //turns on led for 3 seconds
  }
    digitalWrite (led3, LOW);
    pulsePeriod = 0;
  }
}
pulsePeriod = 0;
i = 0;
}
```

Coding Math Combination Word Problems

The following example of a word problem deals with combinations, where each possible scenario is tested (Table 10-1); the inputs and outputs can be displayed in tabular form where the correct solution will easily be seen. As a capstone project, another similar combination problem should be used.

LaTessa and five of her closest friends were celebrating her 16th birthday at a popular vegan restaurant. Each person ordered a salad. Some in the group had a large salad, which was priced at $7.75, and others had a small salad priced at $5.50. The total amount the group paid was $52.00, which included a $10.00 tip to the server. How many people ordered the large salad, and how many ordered the small salad?

It should be understood that there are six persons involved, and that the amount for the food is the grand total for the group, minus the $10.00 gratuity, or $42.00. In prealgebra, the easiest solution is to examine all of the possibilities to find the solution. A similar table might be produced, where the correct combination of large and small orders equates to the total food cost.

Table 10-1. *Word Problem Solution in Tabular Form*

Large	Small	Total
1 = $7.75	5 = $27.50	$35.25 (Too low)
2 = $15.50	4 = $22.00	$37.50 (Too low)
3 = $23.25	3 = $16.50	$39.75 (Too low)
4 = $31.00	2 = $11.00	$42.00 (Correct answer)

Persons with good arithmetic skills might notice the fact that individual costs must add together to produce an even dollar amount, which could then be used in simplifying the procedure for finding the solution.

The programming code in Listing 10-7 provides a solution to the given problem by examining every possibility. Just as a visual learner might produce a table to test the possibilities, a person favoring a sequential logic approach might prefer producing code rather than using a visual aid. In our sample program code running on an Arduino, we use a nested loop method first introduced in Chapter 6 where we generated nonduplicate numbers to be used to simulate a random deck of 52 playing cards. To generate and display a solution to the given problem, the code is written to jump out of the loop and go to a print subroutine when the correct result occurs.

Listing 10-7. Solving the Word Problem with Sequential Logic

```
int ArrayLarge [6];
int ArraySmall [6];
int i;
int j;
float total;
float large;
float small;
const int trigger = 7;
boolean runState;

void setup() {
  pinMode (trigger, INPUT_PULLUP);
  Serial.begin(9600);
  Serial.println ("Momentarily press pin 7 to ground to run");
  Serial.println (" ");
}

void loop() {
  runState = digitalRead (trigger);
  while (runState == LOW) {
    delay (200);
    for (i = 0; i < 6; i++) {
      large = i * 7.75; //large size salad is using the slower
                                    //spinning outer loop
      for (j = 0; j < 6; j++) { //small size salad is using the faster
                          //spinning inner loop
        small = j * 5.50;
        total = large + small;
        if (total == 42.00) {
          goto printOut;      //calls print subroutine
        }
      }
      j = 0;
      total = 0;
    }
printOut:      //print the result to screen subroutine
    Serial.print (i);
    Serial.println (" Large Salads");
    Serial.println (" ");
    Serial.print (j);
```

```
    Serial.println (" Small Salads");
    Serial.println (" ");
    runState = HIGH;
  }
}
```

Our coding example uses a brute force method to check every possibility. Although this might not be the most elegant way to find a solution, it is fairly easy to comprehend for a person favoring sequential logic, and due to the tremendously fast speeds of modern processors, the result will be uncovered within the blink of an eye. Without a computer, this method can be a slow process, especially if a great many possibilities are involved.

One mathematical solution is to represent the problem geometrically with the lines of each equation drawn on a Cartesian coordinate plane. The point where the lines intersect identifies the solution. Additionally, through the use of algebra, the word problem presented represents a system of equations, where the number of diners having each of the two dishes is one of the equations. The other equation in the system is the number of dishes expressed as a coefficient of the large and small salad costs, respectively, being equal to the total food cost.

The simplest equation is $x + y = 6$ where x is the small salad and y is the large salad. The second in the system of equations is $x\,(5.50) + y\,(7.75) = 42$. Using the substitution method and solving for x in the first equation:

$$x = 6 - y$$

Now substituting the expression for x into the second equation we have:

$$(6 - y)(5.50) + (y)\,(7.75) = 42$$

$$33 - 5.50y + 7.75y = 42$$

$$2.25y = 9$$

$$y = 9/2.25$$

$$y = 4$$

Now substituting the number 4 for y into the other equation, we have:

$$x + y = 6$$

$$x = 6 - 4$$

$$x = 2$$

As y is the amount of large vegan salads and x is the amount of small vegan salads, the answer is that there were four large salads and two small salads ordered.

A NOTE TO EDUCATORS

As instructors strive to incorporate different modalities to accommodate the wide range of learning styles, using computer coding is now a possibility. It is my belief that coding is actually a new learning style that I call computer-aided sequential logical reasoning. Even those students who can easier solve a problem by more traditional means will benefit from being exposed to the logical step-by-step process of computer coding.

Appendix

Using and Writing Libraries

The Arduino IDE has quite a few example programs that are very helpful to aid in learning coding. Many of the example programs are linked to libraries embedded within the IDE. You can examine the sample programs contained in the IDE by opening the File menu, and then selecting Examples. On some of the more robust programs, you might notice the pound sign (#) and the term `include` followed by a name. This can be seen in the Wi-Fi example program where `SPI.h` and `WiFi.h` are included. Library files that are included with a sketch add additional functionality to the project. The inclusion of library files allows the programmer to use code that has previously been written to perform a common set of tasks or might be somewhat complex.

Many Arduino library files can be freely downloaded from repository Web sites such as `github.com`. To install the libraries, they can be downloaded as a `.zip` file and saved in the Arduino libraries folder, where they should be renamed as one word without any `.master` designation. Once they are saved in the library folder, they can be easily added (Figure A-1). Depending on the version of IDE, start under the Sketch menu, then select Import Library, and finally Add Library. You then navigate to My Documents and the Arduino library folder, which should contain the folder that you previously downloaded and renamed. That folder must contain at least two files: one with the name and extension `.h`, and one with the name and a file extension of `.cpp`.

© Bob Dukish 2018
B. Dukish, *Coding the Arduino*, https://doi.org/10.1007/978-1-4842-3510-2

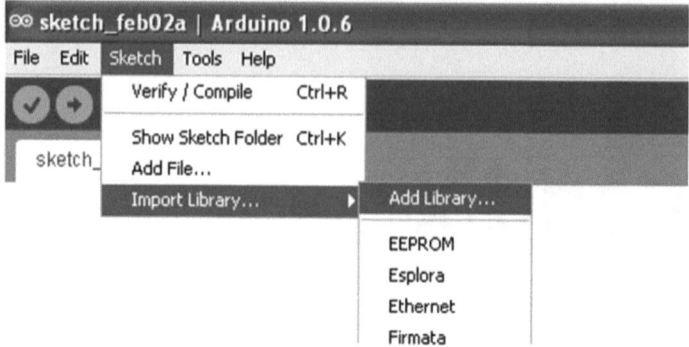

Figure A-1. Adding a library

Once the libriaries are added to the IDE, they can be included into a sketch by clicking the library, or using the pound sign (#) followed by the word include, and then the name of the file with the . h extension. (The . cpp file is linked and does not need to be called for directly.) The downloaded zip folder might also contain a keyword file that is used to highlight the keywords in color as they are used in the sketch. The downloaded folder might also contain example programs that will be displayed in the Include Library section of the IDE.

In the code files that follow (Listings A-1, A-2 and A-3), we present a simple example of how to write a library and how to write a sketch to then use the library functions. In writing the code using the Microsoft Windows Notepad program, or any similar simple text program, we write both the . h and . cpp files, and a keyword file. After writing the library files, save them in a folder inside of the Programs\Arduino\Library folder. Once that is done, the example program gives a serial monitor message in the sketch that tells the user to ground pin 7, which will then be followed by the words hello world generated by the library files.

Listing A-1. The Text File That Is Saved with a `.cpp` Extension

```
//myMessage.cpp file for a library

#include "Arduino.h"
#include "myMessage.h"
int readIn;

myMessage::myMessage(int inPin)
{
pinMode(inPin, INPUT_PULLUP);
_inPin = inPin;
}

void myMessage::message()
{
readIn = digitalRead (_inPin);
if (readIn == LOW){
Serial.println("Hello World");
delay(500);
}
}
```

The .h file:

```
//myMessage.h - library header file

#ifndef myMessage_h //checks for a previous install
#define myMessage_h
#include "Arduino.h" //links Arduino standard functions

class myMessage
{
public:
myMessage(int inPin);
void message();
private:
int _inPin;
};
#endif
```

Listing A-2. The Keyword File

```
//myMessage keywords for library:

myMessage    KEYWORD1 //the space must be a tab
message      KEYWORD2 //the space must be a tab
```

Listing A-3. The Arduino Sketch

```
//using a library
//the library prints the message "Hello World"

#include <myMessage.h>

myMessage myMessage  (7);
//ground pin 7 to run library, you can change the pin

void setup(){
Serial.begin (9600);
Serial.println ("ground pin 7 for a message");
//displays a one-time message
}
void loop (){
  myMessage.message(); //runs the function called "message"
                       //from the library
}
```

Answers to Chapter Review Questions and Projects

Chapter 1

Review Questions

1. True

2. d

3. c

4. rectifier

5. amp, watt

6. positive, negative

7. b

8. Science provides the means of gathering a body of knowledge, whereas technology is the application of knowledge.

9. a

10. formula

Project 1

I = 0.025, or 25 mA expressed in engineering notation.

Chapter 2
Review Questions

1. 0.8 volts, 2 volts

2. Digital signals contain discrete voltage levels and are similar to the function of switching a light bulb on or off. Analog signals have an infinite number of possibilities and can be compared to the operation of a lamp controlled by a dimmer switch.

3. b

4. True

5. Power is the rate of energy consumption. (The word *consumption* is misleading, however, because energy is never really consumed, but is rather converted into another form. Typically, electrical energy is ultimately converted to heat.)

6. b

7. b

8. a

9. Noise in electronics and computer situations is caused by electromagnetic interference (EMI). It can be generated by currents in wires, signals switching on and off, motor operation, and so on. (A subset of EMI is radio frequency interference [RFI].)

10. b

Project 2

I = 0.3 amps. LED characteristics will vary and are listed in the device's data sheet, but LEDs used in electronic projects typically can handle no more than 50 mA. The LED will be damaged and a larger current-limiting resistor should be used in the circuit.

Chapter 3
Review Questions

1. b

2. True

3. d

4. a

5. d

6. a

7. False (It runs it while the condition *is* met.)

8. a

9. hardware, software

10. d

Project 3

A solution is to modify the code shown in Chapter 3.

Chapter 4

Review Questions

1. b

2. d

3. a

4. a

5. a

6. c

7. b

8. b

9. c

10. a

Project 4A

One method to restore the LED duty cycle is to make both of the on and off loop times one-half of their present value. To increase the flash sequence, increase the number from 5 to 7.

Project 4B

One method is to connect one LED circuit to positive voltage, and the other to ground.

Chapter 5

Review Questions

1. 6 seconds (This is because the 2 is the number of milliseconds, and $0.002 \times 3,000 = 6$.)

2. a

3. d

4. True

5. a

6. c

7. b

8. c

9. d (because $1110 = $ decimal $14 = $ hex E)

10. c (because 1001 is a binary representation of decimal 8 and 1)

Project 5

Many possibilities exist to make this modification. It could be possible to use the if conditional statement and the condition equal to ==, or not equal to !=. As suggested, it might be helpful to reference the section from the IDE help menu.

Chapter 6

Review Questions

1. d
2. c
3. d
4. True
5. once
6. a
7. b
8. False
9. An algorithm is a structured process that is used to find a solution to a problem.
10. a

Project 6

Be creative, and remember that the teacher has the red pen!

Chapter 7

Review Questions

1. b

2. c

3. It aids in human understanding of the code.

4. True

5. whole numbers/floating point decimals

6. Arrays can store large data sets with minimal use of variables.

7. c

8. a

9. c

10. a

Project 7

Remember to describe your new variables at the top of the code in the area for global general declarations.

Chapter 8

Review Questions

1. a

2. b

3. c

4. a

5. b

6. c

7. b

8. a

9. b

10. b

Project 8

All ordered phenomena tend to dissolve to a state of entropy (randomness). Just as each conversion from one state of energy to another is not 100% efficient and some amount is lost during each conversion as heat, the ultimate result of electric energy turning to mechanical energy, then to acoustic energy, finally ends as the sound waves distort the surroundings and produce heat energy, which ultimately then increases the entropy of the surroundings as the higher level energy dissipates.

Parts List

- 1 Arduino UNO Rev3
- 1 USB cable
- 1 breadboard
- 1 FET electret microphone
- 1 IR receiver module DFR0094 or equivalent
- 1 IR LEDk to match receiver
- 1 LM386 amplifier
- 1 speaker, dynamic or piezo

- 1 NE555 timer
- 2 seven-segment displays 08MAN72
- 1 photoresistor, cadmium sulfide
- 14 LEDs
- 2 2N3904 transistors
- 2 74LS47 ICs

Capacitors

- 1 470 uF
- 3 10 uF
- 1 0.1 uF

Resistors

- 2 47 K Ohm
- 2 10 K Ohm
- 1 4.7 K Ohm
- 2 2.2 K Ohm
- 2 1 K Ohm
- 2 330 Ohm
- 3 220 Ohm
- 2 120 Ohm
- 1 22 Ohm

Additional

- 1 spool of hook-up wire
- 1 bottle of aspirin or other headache reliever

Index

© Bob Dukish 2018
B. Dukish, *Coding the Arduino*, https://doi.org/10.1007/978-1-4842-3510-2